GREENHOUSE OF THE DINOSAURS

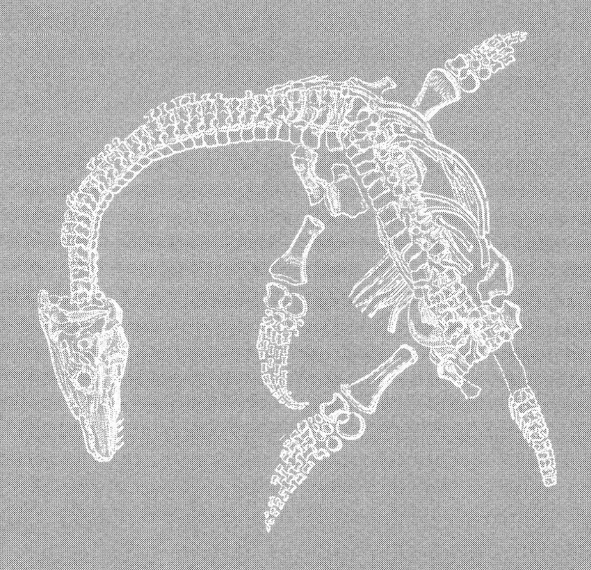

DONALD R. PROTHERO

Greenhouse of the Dinosaurs

EVOLUTION, EXTINCTION, AND THE FUTURE OF OUR PLANET

Columbia University Press *New York*

Columbia University Press
Publishers Since 1893
New York Chichester, West Sussex

Library of Congress Cataloging-in-Publication Data
Prothero, Donald R.
Greenhouse of the dinosaurs : evolution, extinction, and the future of our planet / Donald R. Prothero.; illustrated by Carl Buell.
p. cm.
Includes bibliographical references and index.
ISBN 978-0-231-14660-9 (cloth : alk. paper)
ISBN 978-0-231-51832-1 (e-book)
1. Dinosaurs—Extinction. 2. Climatic changes—Environmental aspects. 3. Geology—United States. 4. Paleontology—United States. I. Title.

QE861.6.E95P76 2009
576.8´4—dc22

2008052555

Columbia University Press books are printed on permanent and durable acid-free paper.
Printed in the United States of America
c 10 9 8 7 6 5 4 3 2 1
Designed by Lisa Hamm
References to Internet Web sites (URLs) were accurate at the time of writing. Neither the author nor Columbia University Press is responsible for URLs that may have expired or changed since the manuscript was prepared.

Frontispiece: Dinosaur Cove in the Cretaceous. (Painting by P. Trusler, used with permission)

To the memory of two great paleontologists
who changed our profession forever:

Malcolm Carnegie McKenna (1930–2008)

and

Stephen Jay Gould (1941–2002)

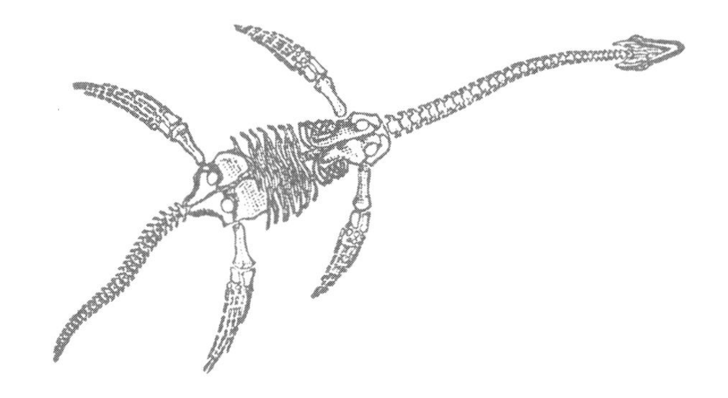

Contents

Preface

May you live in interesting times.
—OLD CHINESE PROVERB

Popular science writing has come a long way since the days of Charles Darwin and Loren Eiseley. Today, the public seems ever more ignorant of science at a time when science literacy is crucial to making public-policy decisions about medicine, the environment, and many other issues. In particular, science is seen as something impersonal and dehumanized, largely due to the scientific tradition of eliminating the observer from the narrative. In scientific publications, the writer must avoid first-person pronouns and use the passive voice, as if the discoveries happened all by themselves, and the human element were unnecessary.

As all good scientists know, however, science is very much a human activity, subject to trends and fads. It is influenced by the culture in which it arises. It is not a cold, impersonal search for what is "out there" beyond the human realm. Science is also a very social activity, from the give and take and camaraderie of professional societies and meetings to the strict gauntlet of peer review to the acrimonious debates over hot topics and controversial ideas. Scientists must go through a long, brutal ordeal known as graduate school to earn their degrees that qualify them for admission into their professions and then through an even longer process in which they must "publish or perish" not only to obtain tenure and job security, but to prove their scientific mettle and keep active by discovering new things that other scientists will consider worthwhile. To me, it is as important to inform the public about this aspect of science as it is to publicize our conclusions and discoveries. In recent years, we've seen a trend for more and more popular science writing to inject this human element and to tell not only the scientific story, but also the social background behind the discoveries, or the writer's own narrative of how he or she was involved in the discovery process. As indicated by sales figures, readers of popular and trade science books apparently enjoy and appreciate this approach because it helps to humanize what is often seen to be a cold and impersonal process.

This book is my own attempt to inject the human side of the profession into the story of the research topics I have had the great fortune to be involved in over the past 40 years. I started out as a kid who loved dinosaurs, but by graduate school I found myself in the center of debates about mass extinctions, tempo and mode of evolution, the geologic timescale, and the great transition from greenhouse of the dinosaurs to our present icehouse planet. I also benefited from interactions with some extraordinary scientists and had the opportunity to work on amazing scientific problems. I was trained at a time of incredible intellectual ferment and even revolutionary transformation in my field. In that respect, I was very fortunate, and I hope that this tale of these "interesting times" will engage the reader as much as my profession has fascinated me.

Acknowledgments

As with any project of this kind, writing and publishing my story would not have been possible without many people's encouragement and support. My editor at Columbia University Press, Patrick Fitzgerald, first urged me to make my previous book, *Evolution: What the Fossils Say and Why It Matters*, more personal and autobiographical, and I have been encouraged by the reactions to that approach ever since. Writers such as Peter Ward, Mike Novacek, Doug Erwin, Bill Schopf, Andy Knoll, and especially Stephen Jay Gould were my models for integrating personal, social, and professional anecdotes into a broader scientific tale. I thank my former professors and mentors in college and graduate school, including Mike Woodburne, Mike Murphy, Dick Tedford, Niles Eldredge, and especially my graduate adviser, Malcolm McKenna. He passed away only months ago, but the influence he had on me and on my entire profession was and still is enormous. Likewise, after seven years I still feel the pain of the loss of Stephen Jay Gould, who not only was a friend and mentor, but also helped me in my career many times when I needed him. I thank Rich Stucky, Dave Bottjer, Tom Rich, Linda Ivany, and Ellen Thomas for reviewing parts of the book, and Patrick Fitzgerald, Meredith Howard, and Marina Petrova at Columbia University Press for their efforts in producing it. I thank Carl Buell for his amazing drawings.

Finally, this project would never have happened without my family's love and support: my parents, Shirley and Cliff Prothero, who encouraged my love of animals and dinosaurs and never urged me to seek a practical career when prospects for paleontology employment were bleak; my wonderful sons, Erik, Zachary, and Gabriel, who make it all worthwhile; and especially my amazing wife, Teresa, who is my strongest supporter and closest friend.

GREENHOUSE OF THE DINOSAURS

Three hypsilophodontid dinosaurs gaze up at the Southern Lights, while an ornithomimid dinosaur sleeps the long, dark winter away in Cretaceous Australia. (Painting by P. Trusler, courtesy T. Rich)

1 | Greenhouse of the Dinosaurs

The heat of European latitudes during the Eocene period ... seem[s] ... equal to that now experienced between the tropics.

—CHARLES LYELL, *PRINCIPLES OF GEOLOGY*

Dinosaurs of the Arctic Jungles

When one thinks of dinosaurs, the first image that pops into mind is that of huge sauropods wandering through warm, lush jungles of conifers and cycads or *Triceratops* and *Tyrannosaurus* rex battling it out in a landscape populated by magnolias and other primitive flowering plants. We hear about the warm climates and dense vegetation of the age of dinosaurs and of the discovery of their remains in tropical and temperate latitudes from Montana to Mongolia to Malawi. Even though Mongolia and Montana are now harsh high-altitude deserts or steppes with blazing hot summers and extremely cold winters, their transformation into the lush landscape of the Mesozoic doesn't seem beyond the realm of possibility. In the past 20 years, however, one of the more astounding discoveries is the revelation that dinosaurs were abundant even in the polar regions above the Arctic and Antarctic circles, where they would have experienced six months of darkness (see the illustration that opens this chapter).

The first evidence of this amazing discovery was accidental and almost completely overlooked. Exploring along the Colville River (figure 1.1) on Alaska's North Slope in 1961, a geologist named R. L. Liscomb was mapping rocks for Shell Oil Company and assessing their oil potential, not looking for fossils. He found and collected some huge bones eroding out of the banks of the Colville River. Not unreasonably, he assumed they were from Ice Age mammals, which are found in abundance in the Arctic region. In 1978, another geologist, R. E. Hunter, found clear dinosaur footprints near Big Lake on the Alaska Peninsula. Finally, in 1984 the legendary paleontologist C. A. "Rep" Repenning of the U.S. Geological Survey (USGS) reexamined the bones collected by Liscomb 23 years earlier and realized they were dinosaur bones. (Rep was a good friend of mine and of most vertebrate paleontologists. In 2005, he was tragically murdered in his own home near Denver by a thief who thought he had commercially valuable fossils in his possession.) The announcement of these dinosaur fossils impelled a number of expeditions by paleontologists and geologists to revisit the Colville site and relocate the dinosaur bones in the 1980s and 1990s, all led by Dr. William Clemens from the Museum of Paleontology at the University of California at Berkeley and Dr. Roland Gangloff of the University of Alaska at Fairbanks. Their discovery was startling—thousands of dinosaur bones (figure 1.2) eroding from the riverbanks for miles along the Colville River, 200 miles north of the Arctic Circle! In 2007, paleontologists even dug a tunnel into the riverbank (figure 1.3) to excavate the bone bed in greater depth and to recover better-preserved bones that had not been shattered by the freezing and thawing of the permafrost near the surface. The Colville River bones came from the end of the age of dinosaurs, the latest Cretaceous period (for the timescale, see figure 1.3), about 69 million years old, and some of the dinosaur tracks there date to the middle part of the Cretaceous, about 90 to 110 million years ago.

The Colville River localities have yielded at least 12 different species of dinosaurs so far, with the most abundant being the huge 13-meter (40-foot),

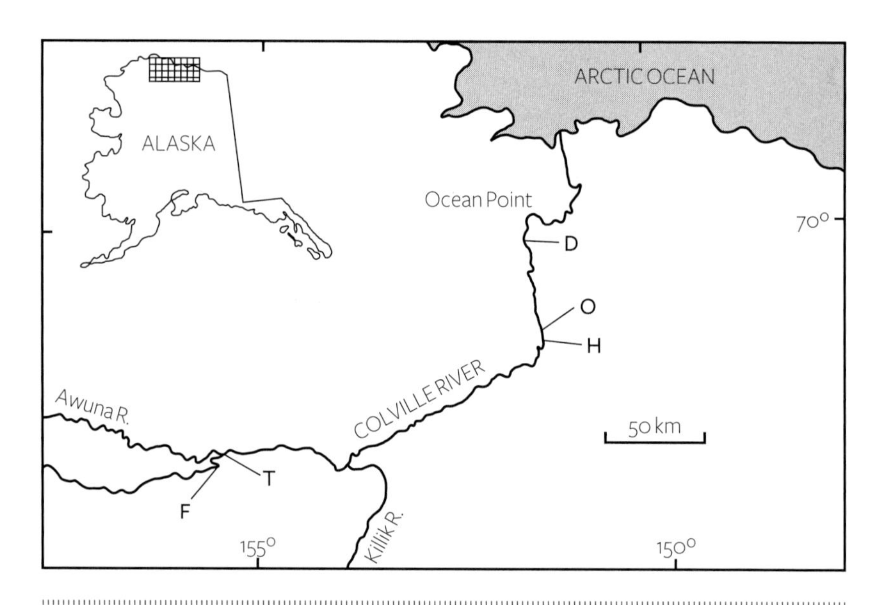

FIGURE 1.1 The Colville River dinosaur localities on the North Slope of Alaska. F = footprint; D = femur; T = turtle and tooth-marked clam; O = occipital condyle; H = horn core. (Redrawn from Parrish et al. 1987)

FIGURE 1.2 In the Liscomb bone bed, (*A*) Kelly May of the University of Alaska Museum of the North excavates a hadrosaur tibia, and (*B*) museum team members excavate hadrosaur material in a permafrost tunnel. (Photographs courtesy K. C. May, University of Alaska Museum of the North)

MILLION YEARS AGO	ERA	PERIOD		EPOCH	MAJOR EVENTS
1.8	CENOZOIC	Quat.		Pleistocene	100,000 yr. ice age cycles
5		Tertiary	Neogene	Pliocene	Panama closes/Arctic freezes *Great American Interchange*
				Miocene	Spread of grasslands *Grazing mammals dominate* Permanent Antarctic ice cap *Mastodonts reach N. America* *African mammals reach Eurasia*
24			Paleogene	Oligocene	Australia/S. America separate Major glaciation/sea-level drop *Grande Coupure*
34				Eocene	First Antarctic glaciers Australia/Antarctica separate Impacts but no extinction Global cooling/mass extinction *Extinction of tropical marine life* *Origin of whales, bats* Maximum greenhouse, poles warm Collision of India/Asia to start Himalayas
55				Paleocene	Global deep-water warming event *Origin of rodents, even- and odd- toed hoofed mammals in Asia* *Evolutionary radiation of mammals*
66	MESOZOIC	CRETACEOUS		Maastrichtian	MASS EXTINCTION Chicxulub impact, Deccan volcanism Sea level drops
				Campanian	Rocky Mountains rise *Peak of duckbill dinosaurs, horned dinosaurs, armored dinosaurs, tyrannosaurs*
				Santonian	Sierra Nevada Mountains
				Coniacian	Peak of global warming, highest sea levels
				Turonian	
				Cenomanian	
120				Albian	*Radiation of flowering plants*
				Aptian	*Earliest placental and marsupial mammals*
				Barremian	Rapid seafloor spreading, global rise in sea level
				Hauterivian	
				Valanginian	
				Berriasian	*Renewed radiation of ammonites*
165		JURASSIC		Tithonian	*Giant sauropods, stegosaurs*

FIGURE 1.3 Mesozoic and Cenozoic timescale. Climatic events are in boldface; biotic, in italics.

three-ton duckbilled dinosaurs known as *Edmontosaurus*. Two other duckbilled dinosaurs, the large-nosed *Kritosaurus* and the large-crested *Lambeosaurus*, are also known from the area. Next in abundance are the horned dinosaurs, or ceratopsians, including *Pachyrhinosaurus*, with its thick, flat horn boss on the nose and broad frill, and the three-horned *Anchiceratops*, which looked vaguely like *Triceratops* but was more primitive. There were also pachycephalosaurs, smaller bipedal dinosaurs that had a thick dome of bone in their skull caps that protected their brain. The function of the thick dome is still controversial, although most paleontologists believe it was used for head-butting combat between males of the same species. A tyrannosaur would certainly make an easy meal of a 5-meter (15-foot) pachycephalosaur with no other armor! Finally, there was the ostrich-like herbivorous dinosaur *Thescelosaurus*, which was about 3 meters (11 feet) long and weighed about 90 kilograms (200 pounds).

The predators include the familiar *T. rex* (known only from a single tooth so far), and a slightly smaller tyrannosaur, *Albertosaurus*, which is better known from southern Alberta (hence its name). There were also the small lightweight predators known as dromaeosaurs, such as *Troodon*, *Dromaeosaurus*, and *Saurornitholestes*, which looked much like the "raptors" of *Jurassic Park* fame. All of these dinosaurs are known from beautiful complete skeletons in the Upper Cretaceous Red Deer River badlands of Alberta, so dinosaurs apparently roamed easily between southern Alberta and northern Alaska in the Late Cretaceous.

Today the Colville River sites are only a short distance from the Arctic Ocean and are frozen over from October to May, with temperatures averaging –16°C (–27°F) in January. Even the brief summer is cold and dry, with highs only about 8°C (46°F) and lows around freezing each night. Nothing grows there today except tundra plants, which are adapted to freezing most of the year and must grow rapidly during the short summer months. Finally, the sites are hundreds of miles north of the Arctic Circle and were so during the Cretaceous as well, so the area experienced four to six months of darkness every year. Clearly, the modern climate and vegetation could not support such a huge diversity of large herbivorous dinosaurs, and today the region has only herds of caribou, musk oxen, and Arctic rodents and rabbits.

Indeed, abundant fossil plant remains are preserved not only in the Cretaceous rocks of the Colville River area, but also in eastern Siberia. Most of the plants were conifers, such as the relatives of the living *Taxodium* (bald cypress), which today are most abundant in the temperate-subtropical wetlands such as the Okefenokee Swamp, the Everglades, and the Mississippi Delta. During the Cretaceous, there were also abundant cycads ("sego palms," which are actually gymnosperms and not true palm trees). These cycads were more vinelike than their modern stumpy palmlike relatives and may have carpeted the landscape. Another gymnosperm, the ginkgo ("maidenhair") tree, was very abundant, as were numbers of smaller flowering plants. Based on their spores, ferns grew in great abundance in the understory of these conifers, as did sphenopsids (scouring rushes or horsetails) in the wetter areas. In short, this plant assemblage was a mixture of swampy vegetation with drier upland forests, not too different from the plants in the warm temperate latitudes of North America today—and nothing like the plants that grow in the modern climate of Alaska.

An analysis of the shapes of the fossil leaves gives a mean annual temperature of 5°C (41°F), with summer temperatures above 10°C (50°F) and winter temperatures at or below freezing (Parrish et al. 1987; Spicer and Parrish 1990). Nevertheless, the tree rings of the logs show strong patterns of seasonal growth, with no growth during the four months of darkness. Nearly all the plants were deciduous and dropped their leaves during the warm, dark winters or died off, as do ferns. So even though the climate was much warmer than it is today, the dark winter landscape would have had almost no fresh green plant food available for the herbivorous dinosaurs to eat.

Paleontologists have long puzzled over how such conditions would permit such a high diversity of dinosaurs. Most think that the dinosaurs migrated up from the south during the summer and back down from the north during the winter. This hypothesis is consistent with the fact that none of the dinosaur bones shows growth rings indicative of slow growth during the winter dormancy. Most of the dinosaurs were large and mobile, and their fossils have been found from Mongolia and China to southern Texas, so they clearly could walk long distances. However, there are also very small dinosaur teeth in the North Slope faunas, suggesting that smaller nonmigratory dinosaurs were also present, or even juvenile dinosaurs that must somehow have shut down their metabolisms and toughed it out for four months of darkness and near starvation.

Dinosaurs of Darkness Down Under

Almost the same story can be told about the Southern Hemisphere in the Cretaceous. Although the Southern Hemisphere dinosaurs are not as abundant as they are in the Arctic, nevertheless there are some important Cretaceous finds here. The first dinosaur fossil finds in Australia were made in 1903 by the pioneering Australian geologist William Hamilton Ferguson, who found a dinosaur claw. Since 1984, my friends Tom Rich of the Victoria Museum in Australia and Patricia Vickers-Rich of Monash University have been working on the problem. Both Tom and Pat were students of my graduate adviser, Malcolm McKenna, at Columbia University. They finished their doctorates in 1973, just a few years before I arrived in the program, and then took positions in Australia. Every austral summer or fall they lead crews to the beach of Inverloch, Victoria, Australia, about 145 kilometers (90 miles) southeast of Melbourne. At a site known as Dinosaur Cove (figure 1.4*A*), they use dynamite and heavy equipment to tunnel into the overlying beach cliff, then rock saws and large hammers to remove the hard blocks of fossiliferous sandstone and chisel out the precious dinosaur fossils. Years of hard labor have produced a great number of dinosaur fossils, including the little hypsilophodontids *Leaellynasaura amicagraphica* (figure 1.4*B*) and *Atlascopcosaurus loadsi*, the small predatory coelurosaur *Timimus hermani*, and ostrichlike oviraptorid dinosaurs as well. Two of these dinosaurs were named for the Riches' children, Leaellyn and Tim.

FIGURE 1.4 Dinosaur Cove in (*A*) the present and (*B*) the Cretaceous, with the carcass of *Leaellynasaura amicagraphica* about to be fossilized. ([*A*] photograph courtesy T. Rich; [*B*] painting by P. Trusler, used with permission)

These fossils are much older in the Cretaceous (95 to 110 million years old) than those from Alaska or Siberia, but they were formed in rocks that were slightly more distant from the South Pole (about 80° south latitude). At that time, Australia was still attached to East Antarctica as part of the great supercontinent Gondwanaland, which didn't completely break up until the Late Cretaceous. *Leaellynasaura* is the most complete of these fragmentary fossils. It was a small bipedal herbivore, less than a meter in length. Its skull shows enlarged eye sockets and huge optic lobes of the brain, suggesting that it had very large eyes and was adept at seeing in low-light conditions. Detailed examination of the bone histology of *Leaellynasaura* shows no obvious growth lines, indicating that these creatures were active year round and did not hibernate or become dormant during the long dark winters when no plants were growing. By contrast, the histology of the little predator *Timimus* does shown pronounced growth lines, suggesting that these animals did become dormant in the winter.

Given these animals' small size, adaptations for the dark conditions, and the fact that the path to warmer climes to the north was blocked by a great inland sea that covered most of Australia during the Cretaceous, it is unlikely that they migrated away during the dark winters, as has been postulated for the Alaskan dinosaurs. Despite the darkness, the world of the Antarctic in the Cretaceous was not barren, although it was probably cold. Geochemical analysis of the bones suggests temperatures ranging from –5°C to 6°C (23 to 42°F, like that of modern Nome, Alaska), although the paleobotanical evidence suggests a slightly warmer summer temperature average of 10°C (about 50°F, like modern London). The landscape was green and lush (see the illustration that opens this chapter), with abundant ferns and *Araucaria* trees (Norfolk Island pines or monkey puzzle trees). There were also abundant ginkgoes, cycads, and podocarps, all common during the Jurassic and Early Cretaceous in lower latitudes as well. At this time, angiosperms, or flowering plants, were just beginning to evolve in lower latitudes, so they are rare at high-latitude localities such as Dinosaur Cove. Although most of the plants were deciduous and dropped their leaves or became dormant during the months of darkness, there were some evergreens as well.

In addition to the dinosaurs, fossils of fish, turtles, flying pterosaurs, birds, and amphibians have also been recovered from the site, so it was a rich locality supporting a full range of cold-blooded and warm-blooded animals not found in freezing climates today. Another slightly younger Cretaceous locality near Inverloch has also produced the tiny lower jaw of a shrew-size mammal known as *Ausktribosphenos nyktos* (Rich et al. 1997), which is more primitive than any living mammal group, including the egg-laying platypus and the pouched marsupials. Indeed, its exact placement within the Mesozoic mammals has long been controversial because its lower teeth seem to be turned backwards relative to any mammal known from the rest of the world. I vividly remember when this specimen was first announced at the Society of Vertebrate Paleontology (SVP) meeting in 1999, and Tom Rich showed it to me. I didn't know *what* to make of it!

Both the North and South Poles were much warmer and more lush with vegetation than they are today and supported a diverse fauna of dinosaurs and other vertebrates that lived in four to six months of darkness every winter. Earth as a whole clearly must have been much warmer for its poles to have so much warmth and vegetation. In addition, there must have been no obstacles for the circulation of tropical oceanic waters toward high latitudes to bathe the polar regions and spread the warmth up to the poles, where there is so little sunlight. Indeed, it is well established that the later half of the Mesozoic was a "greenhouse world" with no polar ice caps and high levels of carbon dioxide in the atmosphere. Very high sea levels drowned most low-lying continents and flooded them with fish, ammonites, and huge marine reptiles. This "greenhouse of the dinosaurs" prevailed from at least the Middle Jurassic (about 175 million years ago) and began to vanish during the Eocene (about 45 million years ago) (see figure 1.3). How this transformation occurred is the subject of the rest of this book.

But what factors can explain the "greenhouse" conditions of the later Mesozoic? Geologists agree that there was much more carbon dioxide in the atmosphere in the Cretaceous than at any time since then—perhaps 2,000 parts per million (ppm)—almost ten times the present value of about 300 ppm, the level of atmospheric carbon dioxide in our atmosphere until recently (that is, until our burning of fossil fuels in the past century triggered the recent rise in greenhouse gases). Researchers have proposed a number of different sources for this excess carbon dioxide in the Cretaceous. Certainly one of the factors was the extraordinarily high rate of seafloor spreading and volcanic eruptions along midocean ridges. These phenomena not only produced new seafloor and oceanic crust, but released enormous volumes of carbon dioxide and other greenhouse gases from the mantle. Many studies have shown that the Cretaceous had some of the fastest seafloor spreading ever seen, as most of the continents that were once united in the Pangea supercontinent began swiftly to pull apart, producing the Atlantic Ocean and separating Africa from Indian from Australia-Antarctica.

Roger Larson of the University of Rhode Island has also pointed to another potential source of the excess carbon dioxide. Deep under the western Pacific Ocean are huge submarine oceanic plateaus that are completely invisible from the surface of the ocean. In fact, they were unknown to science until the 1950s, when oceanographic voyages began routinely to survey and map the ocean floor and discover its surprising and amazing topography. These plateaus are made entirely of basaltic lavas erupted by huge submarine volcanoes during the Cretaceous. The biggest is the Ontong-Java Plateau in the South Pacific, which erupted 1.5 million cubic kilometers (360,000 cubic miles) of lava in only a million years about 122 million years ago. Others include the Hess Plateau and the

Shatsky Rise. Larson points out that the volume of lavas was so immense that it would have released huge amounts of greenhouse gases in the process of eruption. Most geologists think that these huge eruptions were the result of a plume of molten rock coming up from the mantle, punching its way through the oceanic crust, and then erupting on a massive scale. Such plumes are referred to as *superplumes* because they are much larger than the plumes of mantle-derived magma that are currently underneath Hawaii, Iceland, and Yellowstone National Park. Such monstrous eruptions of mantle-derived lavas and gases, both from the midocean ridges and from the superplumes, are unique to the Cretaceous and go a long way to explaining its high levels of atmospheric greenhouse gases.

The Real "Jurassic Park"

Before we get to the end of the "greenhouse of the dinosaurs," however, we should look at Mesozoic faunas that were not at such extreme latitudes. One of the most famous of these faunas is the Morrison Formation, a distinctive Upper Jurassic unit made of purple-, gray-, and maroon-banded mudstones that are widespread in the Rocky Mountains. As early as 1877, Arthur Lakes, who taught at what would become the Colorado School of Mines, found gigantic bones in the mountains just west of Denver near Morrison, Cañon City, and Garden Park, Colorado. They were among the first dinosaurs ever reported from west of the Mississippi, including the first good specimen of *Stegosaurus*, the predator *Allosaurus*, and a number of skeletons of the huge long-necked sauropods *Camarasaurus*, *Diplodocus*, and *Titanosaurus*. Most of these specimens were collected for Othniel C. Marsh at Yale University, who had a small fortune at his disposal to pay for fossils as well as for collectors out in the western territories. After Lakes wrote to Marsh offering his huge bones for sale, Marsh bought them all and then hired Lakes for the princely sum of $125 a year to continue working for him. Then Marsh sent one of his own men, Benjamin Mudge, from the rich fossil beds of western Kansas to help Lakes collect. Marsh's men soon cornered the market on large Jurassic dinosaurs in Colorado. Marsh was attempting to outdo his archrival, Edward Drinker Cope of Philadelphia, who battled him at every turn to collect and describe the most spectacular new finds from out west.

Marsh's men had only begun to collect in Cañon City in July 1877 when they got news from two railroad station agents near Laramie, Wyoming, "Harlow" and "Edwards," that they had huge bones for sale. Marsh sent legendary Kansas collector Samuel Wendell Williston out to Wyoming to find out about the mysterious discovery of huge bones near Como Station on the Union Pacific railway line. When Williston arrived with a check to buy the bones from "Harlow" and "Edwards," he found they could not cash the check. The two men were so secretive that they had written using aliases—their real names were William Harlow Reed and Edward Carlin. Such suspicious, paranoid behavior was typical of the time because Cope and Marsh were notorious for stealing each other's finds and hiring collectors away from each other. Marsh and his men even used code words in their telegrams so that Cope wouldn't learn of their discoveries.

As soon as Williston had scouted the locality, Como Bluff (figures 1.5 and 1.6), he realized that it was far richer than anything in Colorado. Marsh swooped in with men and money to make sure Cope couldn't steal the prize. He tried to hire Carlin and Reed to work for him, but negotiations broke down, and Carlin eventually started working for Cope and opened his own excavations. Meanwhile, Williston and Reed opened quarry after quarry in 1878, and Marsh was soon inundated with huge bones that would form the classic bestiary of the Late Jurassic: the sauropods *Apatosaurus*, *Diplodocus*, and *Barosaurus*; the primitive ornithopods *Camptosaurus*, *Laosaurus*, and *Dryosaurus*; numerous samples of *Stegosaurus*; and several predators, including *Allosaurus*. Marsh's collectors worked year round, even into the winter months, quarrying away in blizzard conditions with the temperatures well below freezing.

FIGURE 1.5 Como Bluff, Wyoming, showing the purple- and brown-banded mudstones of the Morrison Formation. Quarry 9, pictured here, is the source of most of the Mesozoic fossil mammals from the area. (Photograph by the author)

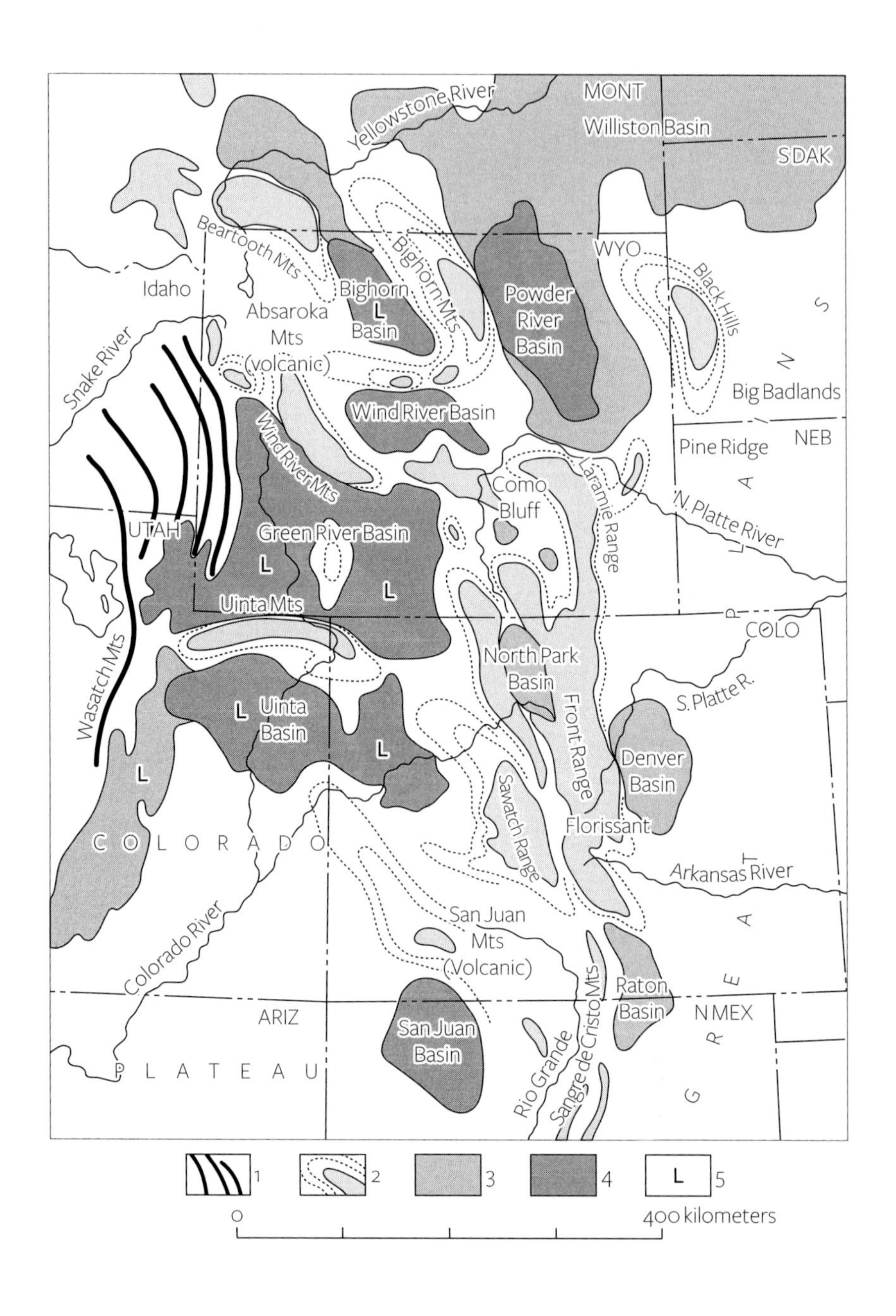

FIGURE 1.6 The Rocky Mountain Cenozoic basins, showing the major localities and regions discussed in this book. *1*, folded and faulted mountains; *2*, Precambrian uplifts; *3*, Paleocene basins; *4*, Eocene basin sediments; *5*, lake deposits. (Modified from King, 1977, Fig. 74).

Marsh's specimens would eventually come to dominate the public understanding of dinosaurs and fossils, and to create an image of the gigantic saurians that we have had ever since then. Dinosaurs had been discovered in England and in New Jersey before the Morrison bonanza, but they were fragmentary, hard to interpret, and seldom mounted or put on display to fire the public imagination. By contrast, many of the Morrison dinosaurs were known from nearly complete skeletons, so they could be mounted in museums and help the world visualize these gigantic creatures. Marsh's "*Brontosaurus*" specimen was one of the first to be mounted in this way (along with a larger "*Brontosaurus*" skeleton that Henry Fairfield Osborn had mounted at the American Museum of Natural History in New York around 1900). Consequently, "*Brontosaurus*" became the most famous of all the dinosaurs, the iconic image of these huge beasts. Ironically, the name "*Brontosaurus*" is no longer valid or used by paleontologists because Marsh himself had already described *Apatosaurus* 1877 from the same quarries at Como Bluff. He then named "*Brontosaurus*" in 1879 from another more complete Como Bluff specimen. They are the same animal, as Elmer Riggs realized in 1903. By the rules of zoological names, the first name used takes precedence over later names, so *Apatosaurus* is the correct name for fossils once called "*Brontosaurus*."

Likewise, Marsh got the head of *Apatosaurus* ("*Brontosaurus*") all wrong as well. His original specimen had no skull attached, so he guessed incorrectly that it had a short-faced, high-domed skull like that of *Camarasaurus*. In the 1970s, however, paleontologists Dave Berman and Jack McIntosh discovered that one *Apatosaurus* specimen did have a decent skull near the end of its neck, and that skull looked more like the long-snouted *Diplodocus*.

For many decades after these great Morrison dinosaurs were first put on display, paleontologists would visualize them as slow, stupid, lumbering lizards floating in swamps, with their tails dragging behind them. But in the 1970s a number of paleontologists began to rethink this old idea. Some argued that these dinosaurs were "warm-blooded" and had high metabolisms like birds or mammals. The "warm-blooded" dinosaur controversy raged for more than a decade, but now seems to be resolved. The smaller predatory dinosaurs (such as the "raptors" of *Jurassic Park* fame) were certainly "warm-blooded" because at their small body size and high levels of activity, they would need a high metabolism to be successful. Indeed, there is good evidence that "raptors" and most predatory dinosaurs (including even *T. rex*) were covered by a downy coat of feathers for insulation, so these animals were not slow and stupid, but active, smart, and warm-blooded.

The size of the huge dinosaurs such as the long-necked sauropods presents a different problem, however. At such large body sizes, these animals had a relatively small surface area compared to their huge volume and no obvious ways of rapidly gaining or losing heat from their bodies. The living elephant is presented with the same dilemma. At its huge size, it must spend much of its time in water or resting in the shade to dump excess body heat. Its huge ears are radiators that shed heat from its body. Most sauropods would have had even greater difficulties if they were "warm-blooded" and generating body heat from metabolism of food. Instead, such large beasts could not use metabolic body heat at all, but kept warm through the warm climates around them. With their large size, they would have gained or lost body heat only very slowly, so they could obtain a stable warm body temperature by sheer size alone. This strategy is known as *inertial homeothermy* or *gigantothermy* and probably characterized all of the larger nonpredatory dinosaurs, including sauropods, stegosaurs, horned dinosaurs, duckbills, and many others.

Thus, we now see all dinosaurs as being much more active and intelligent than we once did. We have evidence from trackways that instead of being slow and sluggish in the swamps, many could move pretty fast. Many dinosaurs, including raptors, duckbills, and sauropods, had specializations in their backbones that enabled them to hold their tails out horizontally like a balancing pole, without ever dragging them on the ground. If you look at a good skeleton of a duckbill dinosaur, you will see the elaborate criss-cross truss-work pattern of ossified tendons in their tail that held the entire structure in a straight, rigid fashion. Likewise, most sauropods had a number of specializations in their neck and tail vertebrae that helped them hold their necks and tails fairly straight out of their bodies, probably in a horizontal position most of the time.

Early paleontologists who collected these Morrison dinosaurs saw the purple-, gray-, and maroon-banded mudstones and the tan river-channel sandstones, and they visualized them as deposits of big, humid swamps. To some extent, this interpretation was biased by the fact that the Morrison was dominated by huge sauropods, and paleontologists thought sauropods were so huge that they needed to be buoyed up by water. More recent studies have negated this old picture, however. Peter Dodson and others (1980) reexamined the sediments of the major Morrison dinosaur localities, and they view the world of the Late Jurassic as a broad floodplain-and-lake system, with seasonal droughts and relatively little standing water. If there had been extensive swamps in the Morrison times, significant coal deposits would have been formed when these swamps turned to stone, but there are none. Dodson and his colleagues (1980) argue that this strongly seasonal floodplain was probably inhabited by migratory herds of large sauropods, which needed to keep moving to find both fresh vegetation after they stripped a region bare and freshwater, which would have dried out locally.

This is the way science operates. Old ideas are constantly challenged, and when new evidence emerges, we need to rethink all the old scenarios. Children's dinosaur books are out of date if they use the name "*Brontosaurus*" or portray these animals as dragging their tails and floating in swamps—but that's the price of progress. Fortunately with the *Jurassic Park* boom in dinosaur paraphernalia, we are quickly seeing a new generation of dinosaur books that incorporate the current ideas.

Kids and "Dino-mania"

Even though my own research has focused largely on the Cenozoic (the "Age of Mammals"), I've had the great privilege of working dinosaur-bearing deposits many times in my career. I'm one of those kids who got hooked on dinosaurs at age four and never grew up. In tenth grade, I already knew where I was going to study paleontology in college, and by the time I reached college I was fully committed to taking every class I could to become a good paleontologist.

I also learned that although dinosaurs were cool, not many new dinosaur fossils were available to work on in the early 1970s. The study of dinosaurs was very crowded with paleontologists competing with each other for the handful of good specimens. By contrast, fossil mammals were just as cool, far more abundant, and better preserved, with many more research opportunities, and the field was nowhere near as overcrowded with other researchers. By my senior year, I was already doing projects in fossil mammals from the Eocene. In my senior year, I was awarded a three-year National Science Foundation (NSF) fellowship, which would pay for my first three years of graduate school no matter where I went. I originally planned to start at the legendary program at Berkeley, the only independent Department of Paleontology in the country at that time (now merged with organismal biology to form the Department of Integrative Biology). I expected to be in Berkeley for two years to earn my M.A., then get my Ph.D. at the even more famous Columbia University program at the American Museum of Natural History in New York. I was accepted to Columbia first and turned it down because I hadn't heard from Berkeley yet. But then my undergrad adviser, Mike Woodburne (himself a Berkeley alumnus), received a call from the Bay Area. The program was not allowed to take *any* new students that year because they had so many Ph.D. students who were still not finished after six to eight years or more, and the administration was clamping down on them. I quickly called Columbia and sheepishly asked my future adviser, Malcolm McKenna, if his offer were still good. Luckily, they had held my slot, and the following fall I moved to New York and discovered the amazing world of the American Museum.

When I arrived, the museum was a stunning revelation. I had seen its famous dinosaur halls on my first visit to New York in the late 1960s, but I never realized that far more fossil material lay in storage in the collections that were open only to legitimate researchers. The public displays were just the tip of the iceberg. In the Frick Wing alone, there were seven floors of fossil mammals (chapter 2), two whole basements of dinosaurs, and another basement with nothing but fossil fish. Even more important was the American Museum's intellectual legacy. The collections included the critical specimens gathered in the 1860s to 1880s by pioneering paleontologist Edward Drinker Cope as well as the crucial collections built up by early-twentieth-century giants of the field Henry Fairfield Osborn, William Diller Matthew, Walter Granger, Barnum Brown, Edwin H. Colbert, and George Gaylord Simpson. Nearly every vertebrate paleontologist alive today can trace their intellectual roots through their graduate advisers and back in time to these men, who trained nearly everyone in the field in the early twentieth century. Even though the American Museum had modern classrooms and labs when I started there, it was also brimming with century-old retired exhibits, models, restorations, and paintings, and even the classroom had a cabinet with the old glass lantern slides that William King Gregory had taught with in the 1920s. The research library included the personal books and journal collections of Osborn and nearly everyone else who had worked there a century earlier, along with Cope's rock hammer and desk.

Even more impressive was the current intellectual state of the American Museum. I was a student of Malcolm McKenna, widely acknowledged as the foremost expert on fossil mammals in the world and a true genius to boot. (He passed away on March 3, 2008, just as I was finishing this book.) Many of the mammalian paleontologists currently in the major museums across the country were his students at one time or another: Bob Emry at the Smithsonian, Bruce MacFadden at Florida, John Flynn (formerly at the Field Museum in Chicago and now Malcolm's successor at the American Museum), Rich Cifelli in Oklahoma, Bob Hunt in Nebraska, as well as many others. In fact, nearly all of Malcolm's highly select group of students (he seldom took more than one a year) went on to success in our profession. No other graduate program can match this track record, where typically only one in ten students who finishes a Ph.D. gets a job.

In addition to Malcolm's expertise on Mesozoic and early Cenozoic mammals, the American Museum also had Richard Tedford on staff, who knew more about later Cenozoic mammals and localities in North America and Australia than any person alive. Eugene Gaffney, a fossil turtle specialist, occupied the fossil reptile position once held by Barnum Brown and then by Edwin Colbert. In the 1980s, the museum added a dinosaur paleontologist, Mark Norell, and another mammalian paleontologist, Mike Novacek (now provost of science). I was one of the last students to be taught by legendary fish paleontologist Bobb Schaeffer, mentor to many of the fossil fish specialists alive today. In short, the museum had not only immense fossil collections, but immense intellectual capital as well. Even the scientific assistants, such as Earl Manning, John Wahlert, and Henry Galiano, were experts on particular groups of fossil mammals. Earl in particular took me under his wing and taught me more than anyone else did. I also learned a great deal from my brilliant officemates and other grad students, including Dan Chure (now the paleontologist at Dinosaur National Monument), Ronn Coldiron (now an elected official in the Silicon Valley), George Engelmann (now at the University of Nebraska, Omaha), Rich Cifelli, and John Flynn.

In addition to hosting amazing fossils and brilliant minds, the American Museum was the site of a revolution in biology and classification as well: *phylogenetic systematics* or *cladistics*. Originally developed by the German entomologist Willi Hennig in the 1950s, cladistics had swept through not only entomology, but also ichthyology and other fields of biology by the late 1960s and early 1970s. It was a radical new way of looking at the classification and phylogeny of organisms wherein the taxonomic groups are based strictly on shared evolutionary specializations and all groups are defined by these evolutionary novelties. Only groups that include a common ancestor and all its descendants are valid, meaning that a traditional group such as "reptiles" is invalid unless it also includes their descendants, the birds.

Cladistics threw out the window all the traditional classification schemes developed over the previous two centuries, but it also solved a much larger number of taxonomic problems. It was soon adopted by nearly every biologist who classified organisms (for a detailed discussion, see Prothero 2003:chap. 4 or Prothero 2007a:128–135). Such a controversial new approach was naturally a highly polarizing influence in a traditional institution such as the American Museum. Originally, only the entomologists and ichthyologists favored cladistics, whereas the mammalogists, ornithologists, and herpetologists were dead set against these ideas. Paleontologists at first rejected some of the more extreme claims (for example, the cladistic assertion that the stratigraphic order of fossils cannot be used to assess their phylogeny), but as time went on, more and more paleontologists came to see the merits of cladistics and adopted it.

Every third Thursday of the month, the American Museum would invite a speaker to its seminar series, the Systematics Discussion Group, where the speaker would be caught in a lion's den between two opposing camps with fundamentally different philosophies about biology, evolution, and classification. At some meetings, each camp's members actually sat grouped together on either side of the central aisle, like the Labor and Tory parties in Parliament, and on occasion the cladists wore T-shirts with the "Willi Hennig, Superstar" logo on them. Each meeting started out relatively quietly as the speaker gave his or her talk, but as soon as the questions began, the verbal battles would break out between the two sides of the room. In some cases, these discussions developed into shouting matches, or the opposing parties would call each other names and blame the other side for lying or character assassination. This was no staid, objective way of doing science, as the popular stereotype suggests. Instead, it is typical of sciences where new and controversial ideas are fighting for attention, and the human side of scientists is very apparent through it.

I had heard none of this controversy during my undergraduate education because the ideas were just then taking root and were confined largely to educational and scientific institutions in New York City. But Earl Manning and my grad student officemates quickly had me unlearn some of the old notions I'd been taught as an undergrad and jump into the new way of thinking because it was taught in every class and was dominating the scientific literature we were reading. Soon we were trying to draw cladograms of the entire Mammalia, an exercise that could be done only at the American Museum with its superb specimens of nearly every group of fossil mammals. The biggest breakthrough occurred in 1975, when Malcolm McKenna published a legendary paper that not only proposed controversial ideas about the homologies of mammalian teeth, but also began the process of classifying all mammals in a cladistic framework. Needless to say, this ambitious and dogma-shattering idea was a shock to paleomammalogists all over the world. They all soon saw the advantages, though. Two decades later, by the time Malcolm actually finished and published his life's magnum opus, a complete cladistic reclassification of mammals (McKenna and Bell 1997), the profession widely accepted this approach.

Needless to say, the late 1970s was an amazing time to be a graduate student: I was learning from the best minds in the profession; both my professors and my fellow students were on the cutting edge of a revolutionary set of ideas that was transforming the fundamentals of paleontology; and I had a huge number of unstudied fossils in the storage floors below me. When I gave my first professional talk at the SVP meeting in Pittsburgh in 1978, I was presenting one of the few cladograms on the entire program. Only a decade later, all systematics talks at the SVP meeting were cladistic, and no one was using the old methods anymore. In addition, the late 1970s brought another revolutionary idea, *vicariance biogeography*, which was closely tied to cladistics and was polarized along the same lines within the scientific community (for further details, see Prothero 2003:chap. 9).

Although from the beginning I was focused on doing research on fossil mammals, opportunities to study and collect dinosaurs nevertheless came up again and again. In the summer of 1977, Malcolm hired me, his first-year student, and John Flynn, his incoming student, as his field crew. We spent much of the summer picking through washed fossil-bearing matrix from various localities, finding tiny bones and teeth, while Malcolm fed and housed us in his amazing ranch just outside Rocky Mountain National Park near Ward, Colorado. But he also took us on a whirlwind tour of all his favorite Mesozoic and early Cenozoic fossil localities in Wyoming, Colorado, and adjacent states, where we saw and collected from nearly every legendary locality that paleomammalogists read about early in their careers: the Bighorn Basin, Teepee Trail Quarry, Togwotee Pass, *Hyopsodus* Hill, Darton's Bluff, Tabernacle Butte, the Green River lake beds with their amazing fossil fish, the Bridger and Washakie basins, the Wind River badlands, Beaver Divide, Bates Hole, Flagstaff Rim, and so on. We visited some important dinosaur-bearing beds that Malcolm had collected over the years, including the Upper Cretaceous Lance Creek Formation in eastern Wyoming, home of *Triceratops*, and the Upper Cretaceous Fox Hills Formation in central Wyoming, which was chock full of the ossified tendons of duckbill dinosaurs.

The most important stop for me, however, was a visit to Como Bluff, Wyoming, home to some of the best-known Jurassic dinosaurs ever found (see figures 1.5 and 1.6). As discussed earlier, Como Bluff was originally collected by O. C. Marsh's crews in the late 1870s and had yielded the first complete skeletons of nearly all the famous Jurassic dinosaurs from the Morrison Formation. Dinosaur bone fragments were so abundant that in one area a local rancher had used them to build the foundation of his house, and so Bone Cabin Quarry became famous. A restored "bone cabin" (figure 1.7) still stands along the highway today. In addition to all these famous dinosaurs, Quarry 9 at Como Bluff had produced a number of tiny jaws of shrew-size Jurassic mammals, the first ever reported from North America. Marsh had briefly described them in the 1887, but George Gaylord

FIGURE 1.7 The modern building made of dinosaur bones from the highway tourist trap near the Bone Cabin Quarry area. Pictured are my 2003 field crew (*left to right*): Matthew Liter, Jingmai O'Connor, Paula Dold, Josh Ludtke, and Francisco Sanchez. (Photograph by the author)

Simpson had done a more complete job as part of his dissertation in the 1920s. In 1968, Tom Rich (then McKenna's student at the American Museum), Chuck Schaff, and Farish Jenkins Jr. (then of Yale University) decided to reopen Quarry 9 at Como Bluff and see if better Jurassic mammals could be found, a full century after Marsh's crew had originally worked there. They headed up a big crew that excavated and handpicked many tons of material to find any of the tiny, pinhead-size Jurassic mammal teeth and jaws. After three hard summers of fieldwork, Tom, Chuck, and their crew had found just a handful of specimens, but they all were of therian mammals (relatives of modern marsupials and placentals), which were supposed to go to Yale according to the agreement worked out before the project. The very few nontherian mammals they had found—members of archaic extinct Mesozoic groups with no descendants—were supposed to go to Tom for his dissertation. Thus, Tom was in a bind. Three years of hard work, and he didn't have enough specimens for a good dissertation. Malcolm helped him out by handing him a project on North American fossil hedgehogs. Malcolm had been planning to work on it for years, and he had all the specimens and literature already assembled, so Tom needed only to plunge in and study them. Thus, he was able to finish his doctorate in time. Tom and Pat were married and soon on their way to jobs in Australia, so the Quarry 9 collection was left unstudied and sitting in Farish Jenkins's office when he moved from Yale to Harvard.

Then Malcolm mentioned during his "Evolution of Mammals" class in 1977 that these Quarry 9 specimens had finally returned to the American Museum after sitting at Yale and Harvard for almost a decade. I was the first to ask if I could study them as my class project. This request soon led to trips up to Harvard to see the rest of the specimens, plus Marsh's Yale specimens (also on loan at Harvard) and stereophotomicrographs of the Late Jurassic mammals of England taken by A. W. "Fuzz" Crompton of Harvard. I eventually saw every Jurassic mammal fossil known from England and North America, so my small class project expanded into a much larger master's thesis and was eventually published in 1981 in the *Bulletin of the American Museum of Natural History*, one of the oldest and most respected museum journals in the world. I was particularly pleased that among the specimens was one new genus and species, which I dubbed *Comotherium richi* in honor of Tom's work (figure 1.8).

While I was up at Harvard, I ran into Richard Estes of San Diego State University. He was then the reigning expert on Mesozoic lizards and other small reptiles and amphibians. I was trying to identify the lizards that Tom Rich, Pat Rich, Chuck Schaff, and crew had collected from Quarry 9, and Dick immediately realized what they were. We wrote up our results, and they were published in 1980 in the British journal *Nature*—my first scientific paper in one of the foremost journals in all of science. I also poked through the rest of Tom's 1968–70 collection and found an ear bone of a Jurassic mammal that I also wrote up and published.

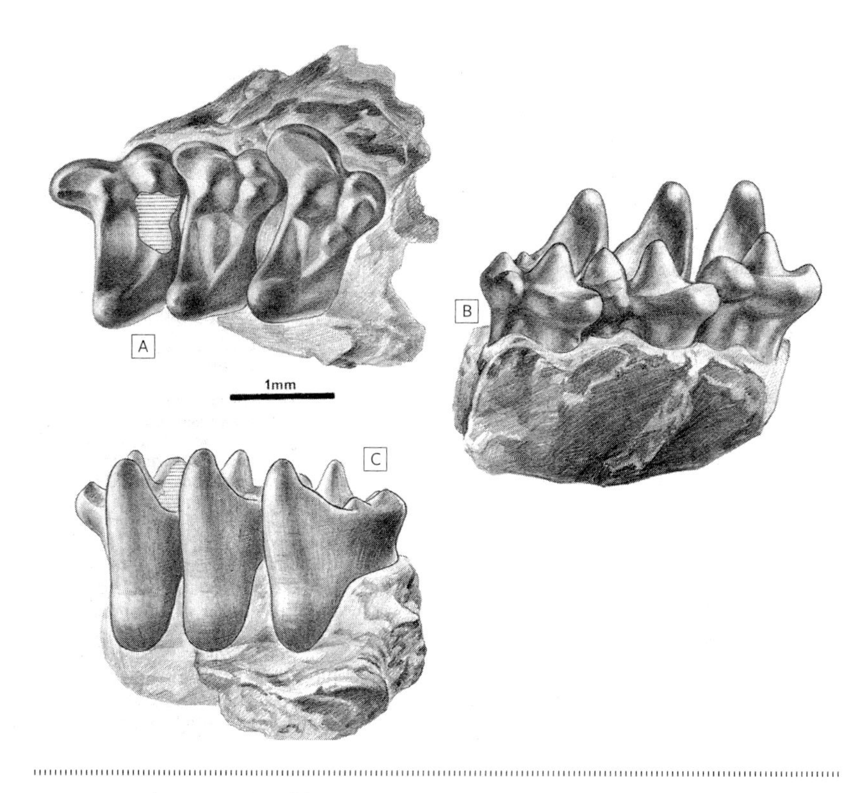

FIGURE 1.8 The upper molars of the Jurassic mammal *Comotherium richi*, a new species I had the privilege of naming in honor of Tom Rich in my master's thesis and first major publication: (*A*) crown view; (*B*) outside view; (*C*) inside view. (Drawing by C. R. Prothero, from Prothero 1981)

A year later, "Dinosaur Jim" Jensen of Brigham Young University approached me to work on an upper arm bone of a Jurassic mammal that he had found in his famous Dry Mesa Quarry in western Colorado, the source of *Ultrasauros*, one of the largest dinosaurs that ever lived. That project too was written up, illustrated with my own hand drawings of the fossil, and published.

After all this research, however, it was clear to me that there was no future for me in Mesozoic mammals. There were no additional specimens to study in the early 1980s, and the field was already crowded with people who were working on what specimens were already available. Plus, the pinhead-size teeth and jaws required a great deal of microscope time to study, and long hours at the microscope were difficult for me with my thick glasses and extreme myopia. So I left Mesozoic mammals behind and moved on to larger mammals, such as rhinos and camels (which don't require a microscope to be seen), and never looked back. Ironically, beginning in the 1990s and continuing now, there has been an explosion of new specimens, new research, and new researchers, so this field is now booming in a way that we could never have imagined when I worked on these fossils in the late 1970s (Kielan-Jaworowska, Cifelli, and Luo 2004). I'm pleased to see the current workers in Mesozoic mammals cite my early work on cladistics of therian mammals and honored when they ask me for my opinion on the matter, but I've moved on and don't work on them anymore.

Since 1980, I have focused primarily on Cenozoic mammals and rocks. In the summer of 1988, however, my field crew and I did some research in the Upper Cretaceous Williams Fork Formation of western Colorado in the Piceance Basin just north of Rangeley. This research was Dave Archibald's ongoing project at San Diego State University, and he invited my crew and me to visit and take paleomagnetic samples (chapter 3). It was a very interesting experience. The rocks looked very different from the Cenozoic beds and were full of coal seams and dinosaur bones. We helped excavate a partial skull of *Triceratops* and saw tyrannosaur bones again and again. The results were analyzed that same summer at the California Institute of Technology (Caltech) lab but still haven't been published twenty years later because of delays by various coauthors!

A Blast of Gas from the Past

Even though the dinosaurs vanished at the end of the Cretaceous (chapter 5), the "greenhouse of the dinosaurs" persisted for at least another 20 million years into the late middle Eocene. In fact, the early Eocene was in many ways warmer and more tropical than even the peak of the Cretaceous "greenhouse" world. Once again, the best evidence comes from fossil vertebrates and plants. The famous lower Eocene beds of the Bighorn and Wind River basins of Wyoming and of the Williston Basin of North Dakota and Montana yield evidence of warm tropical forests full of tree-dwelling and leaf-eating archaic mammals as well as abundant crocodilians and pond turtles.

The Arctic and Antarctic also yield fossil plants that suggest cool temperate conditions even north of the Arctic Circle and south of the Antarctic Circle, where there is darkness for many months of the year. The flora of this region was a paratropical rain forest that included many large conifers (pine trees, the giant *Sequoia*, and the dawn redwood *Metasequoia*), other evergreens, and deciduous trees (elms, oaks, *Liquidambar*, ginkgoes, *Viburnum*, and the bald cypress, *Taxodium*). These forest plants somehow adapted to many months of darkness. However, the temperatures were quite warm above the Arctic Circle in the Eocene, with mean annual temperatures between 19 and 25°C (66 and 77°F) (Wolfe 1980, 1994).

In the 1970s, Malcolm McKenna, Mary Dawson of the Carnegie Museum, and Howard Hutchison of Berkeley led several expeditions to the Canadian Arctic and Greenland to find early Eocene fossils from this region, with amazing success (figure 1.9). During their expeditions to Ellesmere Island in the Canadian High Arctic, they found early Eocene faunas that included abundant crocodilians, monitor lizards, and turtles, as well as gar and bowfin fish, which could not have survived in temperatures that freeze for very long. They discovered a variety of fossil tapirs, primitive horses, rhinos, rhinolike brontotheres (see the drawings that open chapter 4), primates, rodents, fossils related to the living colugos, or "flying lemurs," and smaller mammals, all of which are consistent with the dense forests, abundant fruit fossils, and coal swamps preserved along with them. These mammals show a great similarity to early Eocene faunas of North America and Europe. Indeed, it was such a warm time and so free of polar ice that considerable migration of mammals must have taken place between Europe and North America using Greenland and Iceland as part of the land corridor to cross the North Atlantic, which was much narrower then. When the early Eocene mammals of Europe and North America were first described, paleontologists noted how similar the primates, horses, and many other groups in both places

FIGURE 1.9 Malcolm McKenna collecting fossils from the Eocene beds on Ellesmere Island in the Canadian Arctic. (Photograph courtesy J. Eberle)

were and postulated that there must have been some easy way for them to travel across the Atlantic. But now we can see that the warm polar climates made the corridor across the North Atlantic through Greenland and Iceland much easier to cross in the early Eocene than it would ever be again.

For the best picture of the early Eocene, however, we need to go to the basins of the Rocky Mountains (see figure 1.6): the Bighorn Basin of Wyoming and Montana, the Wind River and Powder River basins of Wyoming, the Williston Basin of Montana and North Dakota, and the San Juan Basin of New Mexico (figure 1.10*A*). These immense bowls filled with thousands of feet of Paleocene and Eocene sediment were warped downward as the Rocky Mountains buckled, folded, and faulted upward starting in the latest Cretaceous. The Rocky Mountain basins provide by far the best terrestrial record of this fascinating time. Nevertheless, the collecting can be challenging. Most of the fossils are isolated teeth and jaws of very small shrew-size to cat-size mammals, so you must collect them on your hands and knees (figure 1.10*B*), with your nose about a foot from the ground, or you will miss them entirely. Over the course of a century, collectors such as Matthew, Simpson, and Granger of the American Museum and Glenn Lowell Jepsen of Princeton University made the tiny fossils of these Paleocene and early Eocene beds their top priority. One of Jepsen's former students, Phil Gingerich (see figure 1.10*B*) of the University of Michigan, has spent almost his entire professional career continuing this collecting effort. He has also trained a whole generation of students in the Bighorn Basin, including Tom Bown (formerly of the USGS), Ken Rose (Johns Hopkins University), Dave Krause (State University of New York at Stony Brook), Scott Wing (Smithsonian), Greg Gunnell and Cathy Badgley (University of Michigan), Jonathan Bloch (University of Florida), Will Clyde (University of New Hampshire), Ross Secord (University of Nebraska), and many others. As a consequence, there are now enormous collections of nearly all the mammals found in the Paleocene and early Eocene of the Rocky Mountain region of North America, and they are among the best-documented and best-studied mammalian fossils in the world.

From these decades of work in the Bighorn Basin and elsewhere, Gingerich and his former students have painted a detailed picture of life in the early Eocene. The barren wastelands of the Bighorn Basin today are hot and dry in the summer and plagued by numbing cold and blizzards in the winter, with a mean annual temperature of only 5°C (41°F) and a spread of more than 33°C (90°F) between daily extremes. In the northern High Plains, it is not unusual for a hot spring or fall day with temperatures higher than 32°C (90°F) to drop below freezing suddenly as an Arctic cold front moves in. In the early Eocene, however, Wyoming, Montana, and North Dakota were mantled by dense tropical forests not too different from those found in Central America today. Mean annual temperatures were as high as 21°C (70°F), and the mean cold-month temperature was only

FIGURE 1.10 (*A*) Panorama of the Paleocene–Eocene beds of the Bighorn Basin, Wyoming. (*B*) Collecting the tiny jaws and teeth from the Paleocene and Eocene requires crawling on your hands and knees, with your eyes just a foot off the ground, or you'll miss the fossils entirely. This crew from the University of Michigan in July 1977 is scouring the ground like a human vacuum cleaner to pick up every single tiny jaw or tooth. The man with the pith helmet in the left background is Phil Gingerich, leader of the group. The woman to his left is Margaret Schoeninger, then a Michigan student and now a professor of anthropology at the University of California, San Diego. The man in the left foreground is Ken Rose, then a Michigan student and now a professor at Johns Hopkins Medical School. The man with the dark shirt on the right behind the stooping student is David Krause, then a Michigan student and now at the State University of New York, Stony Brook. The man in the middle with the visor is John Flynn, then Malcolm McKenna's student at Columbia and now his successor at the American Museum of Natural History. (Photographs by the author)

13°C (55°F). The fossil plants also suggest a very wet climate, with annual rainfall exceeding 1.5 meters (60 inches). The forests formed a multistory canopy like the rain forests do today, with abundant vines and lianas hanging down, perfect for Tarzan to swing on. Many of the fossil plants are from tropical groups that are intolerant of freezing, including citrus, avocado, cashew, and pawpaw trees.

Living in these dense forests was an assemblage of animals that most people would not recognize (figure 1.11). Most familiar would be the abundant crocodilians, pond turtles, and snakes, which love the warmth of the tropics but cannot live in Montana or Wyoming or North Dakota today due to the long, freezing winters. The assemblage of mammals, however, was composed mostly of extinct groups that are unfamiliar to the nonspecialist. They include a variety of very primitive insectivores, lemurlike primates that lived in the tree canopy but also occupied a rodentlike niche as well, and the archaic group of rodentlike, egg-laying mammals known as multituberculates, which were survivors from the Triassic. Down on the ground, nearly all the beasts were archaic hoofed mammals that had teeth suitable for eating a diet of soft, leafy vegetation, along with some mammals that are now extinct and have no modern relatives. The larger predators were mostly from the now archaic extinct group known as creodonts, and true carnivorans (members of the living order that includes dogs, cats, bears, weasels, and seals, among others) were about the size and shape of a raccoon. But there were no really large lion- or bear-size mammalian predators. Instead, that role was taken by huge, 3-meter (9-foot) predatory flightless birds, such as *Diatryma* in North America and *Gastornis* in Europe.

FIGURE 1.11 Diorama of life during the early Eocene. The trees are full of lemur-like primates. On the ground, the huge predatory bird *Diatryma* eats early horses. (Drawing by U. Kikutani)

What could cause the early Eocene world to be so warm, even warmer than it was during the Cretaceous heyday of the "greenhouse of the dinosaurs"? Models of carbon dioxide values place the atmospheric level around 1,000 ppm, about three times the present level, but less than half that of the warmest greenhouse of the Cretaceous (Berner, Lasaga, and Garrels 1983; Berner et al. 2003). Other geologists (Sloan and Rea 1995; DeConto and Pollard 2003) also see high carbon dioxide levels as the chief culprit for these greenhouse conditions. But some paleobotanists have looked at the stomata, the tiny pores found on the undersides of leaves that plants use to exchange oxygen and carbon dioxide with the outside atmosphere (figure 1.12). During conditions of high carbon dioxide, plants make fewer stomata because they can get carbon dioxide for photosynthesis much more easily and lose less water in the process. Conversely, low carbon dioxide conditions trigger leaf growth with more stomata. Dana Royer and colleagues (Royer et al. 2001; Royer 2003) found that the stomatal density on living fossils such as the dawn redwood *Metasequoia* and *Ginkgo* have been fairly constant since the Cretaceous, which suggests that the carbon dioxide level was not that much higher in the early Eocene. Greg Retallack (2001) looked at plant cuticles and found evidence that the carbon dioxide level was only slightly higher than modern levels. Paul Pearson and Martin Palmer (1999, 2000) have demonstrated that the carbon dioxide balance and pH profile in the world's oceans was consistent with an atmospheric carbon dioxide level only slightly higher than we have today.

So if carbon dioxide isn't the chief culprit for explaining early Eocene greenhouse conditions, what is? A number of geologists (e.g., Sloan et al. 1992) point out that methane, "natural gas" or "swamp gas" (CH_4), is also an important greenhouse gas and a much more effective explanation of the early Eocene warming than carbon dioxide is. In 1991, detailed studies were made of the carbon and oxygen chemistry of deep-sea cores that spanned the Paleocene–Eocene boundary about 55 million years ago (Kennett and Stott 1991). It had long been known that at this time there was a major extinction in the benthic foraminifera, the microscopic organisms that lived on the sea bottom, but otherwise there was no evidence of a big mass extinction in other organisms. But these new high-resolution cores showed the interval in great detail and clarity. It was soon apparent that the extinction and the sudden change in ocean chemistry was much more abrupt than previously suspected. Some estimates suggest that the change took place in much less than 10,000 years and that average temperatures in the world ocean abruptly rose by 4°C (about 8°F). Such a rapid and extreme change in ocean chemistry and temperature could not be explained by gradual movement of continents or rearrangement of oceanic currents. Instead, another source was needed.

In 1993, oceanographers discovered that there were huge amounts of methane frozen in little ice cages known as *methane hydrates* or *methane clathrates*

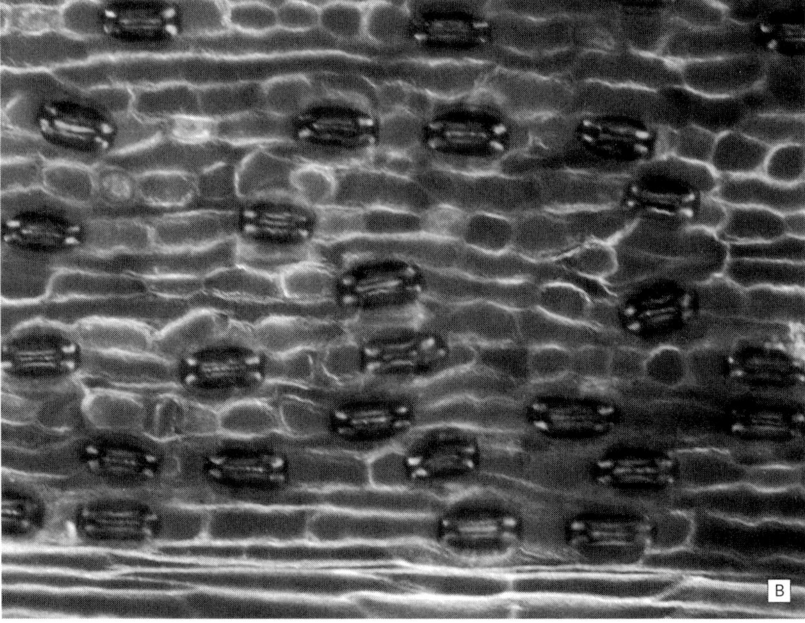

FIGURE 1.12 The stomata of the undersides of the leaves of the dawn redwood (*Metasequoia*): (*A*) in an Eocene example, the clear oval-shaped cells are the stomata; (*B*) modern *Metasequoia*. The density of the stomata is slightly less in the Eocene than in the modern example, suggesting that the atmosphere was richer in carbon dioxide during the Eocene than it is today. (Photographs courtesy G. Doria)

(Kvenvolden 1993). These chemicals were naturally trapped in the pore water of oceanic sediments, where they remained frozen as long as ocean bottom water temperatures were below 5°C (41°F). They occur in huge volumes today in the continental margins, estimated at about 11 trillion trillion grams (about 500 billion trillion pounds), all trapped in a semistable state as long as the oceans remain cold enough. But with the warming during the Paleocene, this frozen methane would have melted and been suddenly released into the world's oceans, killing mostly sea-bottom-dwelling benthic organisms, then causing the planet's oceans and atmospheres to be saturated with natural gas and driving the climate into a "supergreenhouse." Since 1995, geologists have visualized this scenario for the Paleocene–Eocene boundary (Dickens et al. 1995, 1997; Thomas and Shackleton 1996; Dickens, Castillo, and Walker 1998; Thomas et al. 2002). This "blast of gas from the past" apparently warmed the planet so fast that most bottom-dwelling marine life could not survive, and the shallow marine and terrestrial realms experienced record warmth and plant growth. Although the big excess of methane was gradually reabsorbed and oxidized to carbon dioxide, its effects were present for much of the early Eocene.

This "methane burp" was felt even on land. Not only did it trigger global warming and intense tropical plant growth almost to the poles, but it can be seen in the chemistry of the teeth of early Eocene mammals and even in the soils. Their chemical composition shows the same abrupt shift to light carbon-12 that we see in marine sediments and fossils, so this event was global. In fact, this chemical isotope signal is universal at the Paleocene–Eocene boundary, so the boundary is pegged to this marker rather than to some local change in the fossils. And right after the carbon isotope spike is when we see the maximum migration of mammals back and forth between Eurasia and North America. For example, another marker of the early Eocene in the Bighorn Basin and elsewhere in North America is the first appearance of primitive horses, rhinos and tapirs (perissodactyls), even-toed hoofed mammals (artiodactyls), advanced lemurlike primates (adapids and omomyids), and (slightly earlier) rodents, all from more primitive relatives in Eurasia. Thus, the "blast of gas from the past" is the final piece of the puzzle, explaining not only why the early Eocene polar regions were so warm and such a good corridor for mammal migration, but also how these animals managed to cross the North Atlantic across Greenland and Iceland for the first and last time.

I vividly remember many conversations with Malcolm McKenna about this very topic because it was central to work he had been doing since his undergrad days. He was the pioneer who first worked on the early Eocene mammals of northwestern Colorado for his dissertation in the 1950s, found the Arctic mammal fossils, explored Greenland for Eocene mammals, and spent a great deal of time puzzling over the incredible similarities between early Eocene mammals from Europe and those from North America, and over how they might have traveled across the North Atlantic. I'm sure he was very pleased in his later years to hear about all these past developments, which have added a final piece to this amazing puzzle.

Further Reading

Aubry, M.-P., S. G. Lucas, and W. A. Berggren, eds. 1998. *Late Paleocene–Early Eocene Climatic and Biotic Events in the Marine and Terrestrial Records*. New York: Columbia University Press.

Dickens, G. R., J. R. O'Neill, D. K. Rea, and R. M. Owen. 1995. Dissociation of oceanic methane hydrate as a cause for the carbon isotope excursion at the end of the Paleocene. *Paleoceanography* 10:965–971.

Kielan-Jaworowska, Z., R. L. Cifelli, and Z.-X. Luo. 2003. *Mammals from the Age of Dinosaurs*. New York: Columbia University Press.

Ostrom, J. M., and J. H. McIntosh. 2000. *Marsh's Dinosaurs: The Collections from Como Bluff*. New Haven, Conn.: Yale University Press.

Parrish, J. M., J. T. Parrish, J. H. Hutchison, and R. A. Spicer. 1987. Late Cretaceous vertebrate fossils from the North Slope of Alaska and implications for dinosaur ecology. *Palaios* 2:377–389.

Prothero, D. R. 2003. *Bringing Fossils to Life: An Introduction to Paleobiology*. 2d ed. Boston: McGraw-Hill.

Prothero, D. R. 2006. *After the Dinosaurs: The Age of Mammals*. Bloomington: Indiana University Press.

Rich, T. H., and P. V. Rich. 2000. *Dinosaurs of Darkness*. Bloomington: Indiana University Press.

Skelton, P. 2003. *The Cretaceous World*. Cambridge: Cambridge University Press.

Thomas, D. J., J. C. Zachos, T. J. Bralower, E. Thomas, and S. Boharty. 2002. Warming the fuel for the fire: Evidence for thermal dissociation of methane hydrate during the Paleocene–Eocene thermal maximum. *Geology* 30:1067–1070.

Ward, P. D. 2007. *Under a Green Sky: Global Warming, the Mass Extinctions of the Past, and What They Can Tell Us About Our Own Future*. Washington, D.C.: Smithsonian Books.

Wing, S. L., P. D. Gingerich, B. Schmitz, and E. Thomas, eds. 2003. *Causes and Consequences of Globally Warm Climates in the Early Paleocene*. Geological Society of America Special Paper no. 369. Boulder, Colo.: Geological Society of America.

Panorama of the Big Badlands looking east from the top of Sheep Mountain Table in the western end of the Big Badlands. (Photograph by the author)

2 | Bad Lands, Good Fossils

> I was totally unprepared for that revelation called the Dakota Bad Lands. What I saw gave me an indescribable sense of mysterious elsewhere.
>
> —FRANK LLOYD WRIGHT, IN A LETTER TO A FRIEND FOLLOWING HIS 1935 VISIT TO THE BADLANDS

Mauvaises Terres

The heat is unbelievable. The ground is literally too hot to touch or even sit on. Even my boot soles feel as if they're melting. The narrow canyons with their tan and white walls act as a reflector oven, making the air temperatures higher than 43°C (110°F). My eyes are almost blinded by the intense glare, but I dare not wear sunglasses because my quarry is small scraps of bone and shiny black tooth enamel that would be hard to see with sunglasses. These specimens are the crucial clues to finding precious fossils in this hellish, unforgiving place. A large broad-brimmed hat is essential to keeping the intense sun off my neck and face and lots of sunscreen only keeps me from burning to a crisp on the first day. An eager young student at this time in the 1970s, I foolishly don't wear much sunscreen because no one makes such a big deal about the risks of sun exposure and skin cancer. I burn and peel several times each summer. (Now I take no chances, and slather that SPF 50 on really thick before I go out collecting.)

My graduate adviser, fellow students, and I work our way slowly across the rocky ground, stooped over, our faces less than two feet above the ground (see figure 1.10*B*). We're looking for fossils that are often hard to distinguish from ordinary rocks that litter the ground. Here's another shiny cobble that resembles a tooth—nope, just a shiny lump of rusty iron oxide, or hematite. Here's another rock that has a bonelike shape—nope, on closer inspection its internal structure doesn't have that characteristic spongy texture of bone, so it's just a concretion. I am young and naive and proud of my agility at that age, so I clamber up one ridge and along the spine of another, trying to reach places where no one has ever looked before. After a while, I realize that I'm not finding anything because I'm too busy climbing and watching my precarious footing, but my grad adviser has much better luck. He wisely walks slowly along the ravine bottoms, where he doesn't have to worry about falling off a cliff. He is looking for fragments that have washed down from above and accumulated in the arroyo, and if he finds something interesting, he can trace it up the slope to its source.

These false alarms and close calls regarding things that look like important fossils but aren't continue hour after hour, and no one has found even a fragment of something interesting. We've been at it for days, with only a few isolated teeth and bone scraps to show for all our efforts of thousands of hours in the blazing sun. This is the nature of what paleontologists do—finding fossils is very hard and often dangerous work, often with little or no luck despite hundreds of days of work in very promising places where fossils *should* be.

The effort's not that profitable, either. Most professional paleontologists get paid very little for their work as professors or museum curators. To earn a Ph.D. in paleontology, you need to survive at least 10 years of college, many more than a lawyer or an MBA has to endure. That's as much college education as a surgeon suffers through—yet we're paid far less than members of these other professions. Doing paleontology for a living is *not* a way to get rich! It is often jokingly said that there are only two kinds of paleontologists—the rich and dedicated (the wealthy few who can do anything they want with their lives) and the poor and dedicated (the rest of us)—but no one becomes a paleontologist as a way to make big bucks.

The rewards are not financial, but rather intellectual and emotional: the satisfaction of finding something that is new and different and never before seen by human eyes, something that has been buried in the earth for millions of years. Most of the time, the fossils we find are not that important or even identifiable. Even a really good specimen is usually just a duplicate of something already in a museum collection somewhere and often not as good as what we already have. It's valuable to have more specimens of that fossil to increase our sample size and study things such as variability, but an individual specimen is usually not important enough to be published all by itself.

Every once in a while, however, a paleontologist gets really lucky and finds something that is genuinely new to science. It may be years before the specimen is finally cleaned up, prepared, and ready to study, and even longer before it is described and published so the rest of the world knows about it. Talk about slow! Plus, two to four years may elapse from the time the paleontologist writes the paper until the time it's finally published! But when it finally does appear in print, it's a source of great pride to the author, almost comparable to becoming a parent for the first time. I've found my share of new fossils, although I usually end up describing fossils that someone else collected, put in a museum long ago, and didn't know what he or she had. It usually falls to a specialist who visits museum after museum over many years to recognize which fossils in the collections are new, important, and need to be described and published.

Nevertheless, these great discoveries come at the cost of many hard years in the field, working in horrible conditions of heat and aridity, primarily in the desert regions around the world. The best place to find decent fossils are typically deserts because they have little vegetation to cover and destroy the fossils, but lots of naked rock that erodes quickly during rare desert thunderstorms and flash floods.

The early pioneers often named such inhospitable places, and their names reflect their abhorrence of them: Devil's Punchbowl, Devil's Graveyard, Hell's Half Acre, Hell Creek, Purgatory Hill, and other such diabolical and hellish combinations. Ironically, to a geologist or paleontologist, such a name on a map is like a magnet. It is a sure-fire clue that the feature itself is geologically interesting because bare rock is what a geologist or paleontologist needs to see Earth's crust exposed or to find fossils eroding from the naked rock outcrop. By contrast, paleontology and geology in well-watered places with vegetation can be very difficult. Most of the countryside is covered with plants or soils, and the only outcrops are road cuts, quarries, and occasional stream banks and beach cliffs that expose bedrock for geologists to study and interpret. It can be hard to find fossils even in outcrops because road crews try to stabilize and conceal them with Hydromulch, ice plants, kudzu, and other forms of cover that prevent rapid erosion, but also hide all the interesting geology.

The best exposures are where soft silts and sands from million-year-old floodplains and riverbeds have been hardened into stone, then uplifted and eroded down. These erosional features are called "badlands" because to most of the early settlers who pioneered a region, they were useless for growing crops and a bad place to lose a cow. Even the Native Americans stayed away from them, and such sites often became hideouts for bandits and outlaws, where no one in his right mind would want to go unless he was running from the posse—or finding fossils.

The prototype of all the "badlands" around the world is the Big Badlands of western South Dakota (figure 2.1), which are now mostly part of Badlands National Park as well as several Lakota (Sioux) reservations. (Any topography with lots of sinuous bare ridges and canyons can be called "badlands," but "Badlands" with a capital B refers to the Big Badlands of South Dakota.) The Lakota called the place *mako sica* ("bad lands" in Lakota), and early French trappers called it *les mauvaises terres a traverser*, meaning "bad lands to travel across." Early American trappers and traders soon followed the French and relayed their own descriptions. The earliest known written account of the area is that of James Clyman, a mountain man in Jedediah Smith's party who passed through the area in 1823.

FIGURE 2.1 Artist's portrayal of the Big Badlands from the Owen Survey report (Owen 1852).

Clyman wrote:

> a tract of county whare no vegetation of any kind existed beeing worn into knobs and gullies and extremely uneven a loose grayish coloured soil verry soluble in water running thick as it could move of a pale whitish coular and remarkably adhesive[;] there [came] on a misty rain while we were in this pile of ashes [badlands west of the South Fork of the Cheyenne River] and it loded down our horses feet . . . in great lumps[;] it looked a little remarkable that not a foot of level land could be found the narrow revines going in all manner of directions and the cobble mound[s] of a regular taper from top to bottom all of them of the percise same angle and the tops sharp[;] the whole of this region is moveing to the Misourie River as fast as rain and thawing of Snow can carry it. (from Clamp 1928:24, cited in Badlands Natural History Association 1968, spelling as given in the original; see also Clyman 1984)

Aroused by the reports of fossils from these early explorers, a number of early scientific expeditions reached the region by the 1840s. In 1843, Alexander Culbertson passed through the region and made a small collection of mammal fossils that were eventually described by Joseph Leidy, a Philadelphia surgeon and the first vertebrate paleontologist in America. Leidy named the most common

Badlands mammal, the oreodont *Merycoidodon culbertsoni*, after him. In 1846, Hiram Prout of St. Louis described a lower jaw fragment of a huge rhinolike mammal we now know as a brontothere or titanothere. By 1849, geologist David Dale Owens sent his assistant, John Evans, to the Big Badlands, along with a crew of French travelers, cooks, mule drivers, and other support crew. A French traveler in the group, Émile de Girardin, was a mercenary soldier serving as an artist. His account appeared in 1864 in a French travel magazine, *Le Tour du Monde*. He climbed a hill about 100 meters (330 feet) high and saw "the strangest and most incomprehensible view": "At the horizon, at the end of an immense plain and tinted rose by the reflection of the setting sun a city in ruins, appears to us, an immense city surrounded by walls and bulwarks, filled by a palace crowned with gigantic domes and monuments of the most fantastic and bizarre architecture. At intervals on a soil white as snow rise embattled chateaus of brick red, pyramids with their sharp-pointed summits topped with shapeless masses which seem to rock in the wind, a pillar of a hundred meters rises in the midst of this chaos of ruins like a gigantic lighthouse" (De Girardin 1936:60).

De Girardin was amazed by the fossils the group saw in the Badlands. "The soil is formed here and there of a thick bed of petrified bones, sometimes in a state perfectly preserved, sometimes broken and reduced to dust." They found "petrified turtles," some of which were "admirably preserved and weighing up to 150 pounds," and "a head of a rhinoceros equally petrified, and the jawbone of a dog or wolf of a special kind, furnished with all its teeth." In other areas, there were "heaps of teeth and scraps of broken jawbones; . . . bones and vertebrae of the oreodon, the mastdon [*sic*] and the elephant" (De Girardin 1936:60)

Evans himself was impressed not only by the scenic qualities of the Badlands, but by the scientific importance of the region as well. He wrote:

> After leaving the locality on Sage Creek, affording the above-mentioned fossils, crossing that stream, and proceeding in the direction of White River, about twelve or fifteen miles, the formation of the Mauvaises Terres proper bursts into view, disclosing as here depicted, one of the most extraordinary and picturesque sights that can be found in the whole Missouri country.
>
> From the high prairies, that rise in the background, by a series of terraces or benches, towards the spurs of the Rocky Mountains, the traveller looks down into an extensive valley, that may be said to constitute a world of its own, and which appears to have been formed, partly by an extensive vertical fault, partly by the long-continued influence of the scooping action of denudation.
>
> The width of this valley may be about thirty miles, and its whole length about ninety, as it stretches away westwardly, towards the base of the gloomy and dark range of mountains known as the Black Hills. Its most depressed portion, three hundred feet below the general level of the surrounding

country, is clothed with scanty grasses, and covered by a soil similar to that of the higher ground.

To the surrounding country, however, the Mauvaises Terres present the most striking contrast. From the uniform, monotonous, open prairie, the traveller suddenly descends, one or two hundred feet, into a valley that looks as if it had sunk away from the surrounding world; leaving standing, all over it, thousands of abrupt, irregular, prismatic, and columnar masses, frequently capped with irregular pyramids, and stretching up to a height of from one to two hundred feet, or more.

So thickly are these natural towers studded over the surface of this extraordinary region, that the traveller threads his way through deep, confined, labyrinthine passages, not unlike the narrow, irregular streets and lanes of some quaint old town of the European Continent. Viewed in the distance, indeed, these rocky piles, in their endless succession, assume the appearance of massive, artificial structures, decked out with all the accessories of buttress and turret, arched doorway and clustered shaft, pinnacle, and finial, and tapering spire.

One might almost imagine oneself approaching some magnificent city of the dead, where the labour and the genius of forgotten nations had left behind them a multitude of monuments of art and skill. (quoted in Owen 1852:196–197, cited in Badlands Natural History Association 1968)

Evans was awestruck by the richness of the fossil beds:

At every step, objects of the highest interest present themselves. Embedded in the debris, lie strewn, in the greatest profusion, organic relics of extinct animals. All speak of a vast freshwater deposit of the early Tertiary Period, and disclose the former existence of most remarkable races, that roamed about in bygone ages high up in the Valley of the Missouri, towards the sources of its western tributaries; where now pastures the big-horned *Ovis montana*, the shaggy buffalo or American bison, and the elegant and slenderly-constructed antelope.

Every specimen as yet brought from the Bad Lands, proves to be of species that became exterminated before the mammoth and mastodon lived, and differ in their specific character, not alone from all living animals, but also from all fossils obtained even from cotemporaneous [*sic*] geological formations elsewhere. (quoted in Owen 1852:197–198, cited in Badlands Natural History Association 1968)

By the middle 1850s, Joseph Leidy had described a number of important fossils, including the first specimen of an American camel, *Poebrotherium wilsoni* (1847); the first fossil of an American horse, *Mesohippus bairdi* (originally called "*Palaeotherium*" *bairdi* in 1850); and the first American rhino, *Subhyracodon occidentalis* (which Leidy had originally called *Rhinoceros occidentalis* in 1850). He published on many more Badlands mammals in several more books and papers in the 1860s before leaving paleontology altogether in the early 1870s when more aggressive and spendthrift scientists such as O. C. Marsh of Yale and Edward Drinker Cope of Philadelphia came along and cut off his supply of fossils from the West. Nevertheless, the nineteenth and twentieth centuries saw many expeditions to the South Dakota Badlands, so that beautiful complete skeletons of many of these fossil mammals can be seen in museums and rock shops all over the country.

The White River badlands of South Dakota, Nebraska, Wyoming, Colorado, and North Dakota are still among the most fossiliferous deposits anywhere in the world. More than 160 years of collecting have now made the Badlands fossils among the best known and studied in the entire fossil record. They are also among the most commonly poached and collected illegally by commercial collectors, who then place them on the market to fetch astounding prices. Sadly, the immense areas of these badlands cannot be policed by the limited resources of the National Park Service and the Lakota police, so the poachers come and go onto federal or tribal land and steal specimens with impunity. They occasionally get caught and face fines and prison time, but by and large little can be done about such poaching without a huge investment in policing.

Badlands Bestiary

The mammals of the Badlands startled the world in the 1850s, and they are still truly amazing to behold. The largest and most impressive are the huge brontotheres, or titanotheres, which reached elephantine proportions (figure 2.2). Some places in the lowest Badlands unit, the upper Eocene Chadron Formation, are graveyards full of the skulls and bones of these amazing creatures and exploited by museums for more than a century. In most ways, a brontothere might remind you of a large rhino, except that instead of a pointed horn made of compressed hairs (as in a true rhino), brontotheres had huge, blunt, "battering-ram" horns made of bone, which forked into a "Y" shape. The rhinolike shape of these beasts is a clue that they are distantly related to rhinos, horses, and tapirs, not to elephants (despite their size).

At first, scientists envisioned these animals as battling in head-to-head combat like two bighorn rams colliding during mating season. But subsequent work has shown that their skulls are rather lightly built with spongy bone behind the horns, so they could not have endured any heavy impacts without shattering the skull. Instead, brontotheres probably practiced less-violent sparring techniques,

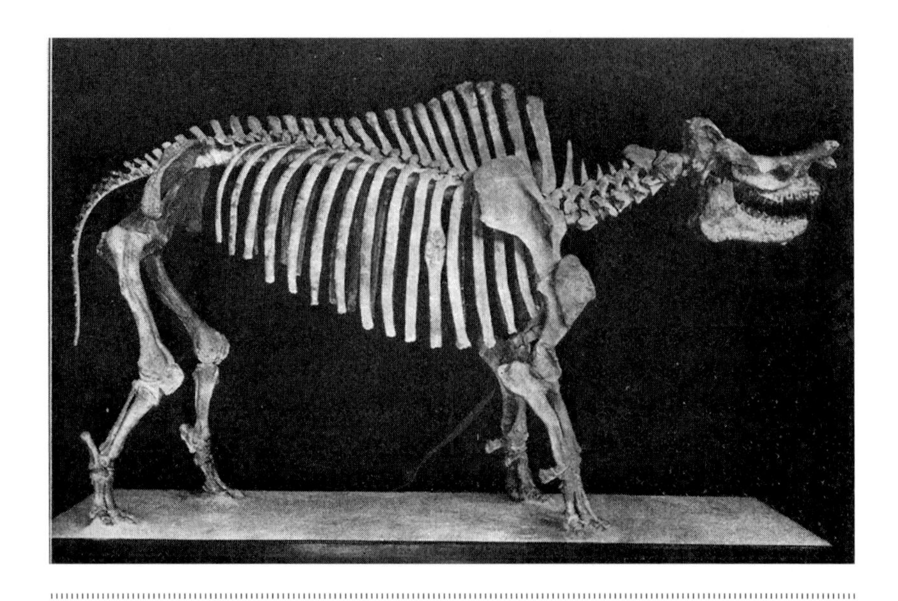

FIGURE 2.2 Skeleton mount of the brontothere *Megacerops* in the American Museum of Natural History, showing the broken and healed rib in the middle of the ribcage.

with some head shoving and slamming of their horns against the flanks of their foes. Indeed, a brontothere skeleton in the American Museum of Natural History in New York has a healed fracture of the ribs in the central ribcage, strongly suggesting a wound caused by just this kind of indirect combat (see figure 2.2).

Brontotheres are a fascinating group, and their evolution from small creatures barely distinguishable from the earliest horses, rhinos, and tapirs about 55 million years ago is a remarkable example of evolution (see the illustration that opens chapter 4). Their story was told at great length by the famous paleontologist Henry Fairfield Osborn, who made the study of brontotheres one of his life's quests. Scion of a rich and powerful family, Osborn at some point met the Philadelphia paleontologist Edward Drinker Cope, who interested him in paleontology. After a Princeton degree, a postgraduate education in Europe, and a stint teaching at Princeton, Osborn set up shop at the American Museum in New York, which he soon made into the foremost museum in the United States. It was he who commissioned its grand halls and raised money from his rich friends, which also paid for amazing scientific expeditions to the American West, Mongolia, and many other places.

Unfortunately, Osborn was also known for his pompous attitude and huge ego as well as for his unorthodox ideas about evolution that none of his peers accepted (Rainger 2004). He culminated his work on brontotheres with the gigantic, two-volume, 1,000-page monograph *The Titanotheres of Ancient Wyoming, Dakota, and Nebraska* (1929). This immense work was so intimidating that no one tried to redo it for almost 60 years. Yet even in Osborn's time his paleontologist colleagues realized that it was badly done. Osborn named almost every decent skull as a distinct species and recognized dozens of different lineages of brontotheres that he claimed evolved in parallel, driven by the mysterious forces of what he called "orthogenesis" and "aristogenesis."

In recent years, Bryn Mader (1989) and now Matthew Mihlbachler (2008) have undertaken the Herculean task of revision, so most of Osborn's invalid species and genera are now abandoned. The major loose end is the final stage of the brontotheres' evolution, when they reached their largest size in the Chadronian (late Eocene). Although the museums around the United States have many excellent skulls, the crucial evidence needs to come from really large collections of brontotheres with excellent field data showing exactly what stratigraphic level they came from. We need many specimens from deposits of the same age (thus showing how much their horns vary within a population) and from collections that show precisely how they changed within the 3 million years of the Chadronian (34 to 37 million years ago). Such a collection does indeed exist, but the fossils are still in the plaster field jackets from when they were collected in the 1940s and 1950s. No one at the American Museum has the interest or funding to have the skulls prepared and cleaned up for study. Several scientists have run into this "plaster wall" and have had to stop studying Chadronian brontotheres until the specimens can be adequately prepared. While Scott Wing was a Yale student, he came to the American Museum gung-ho to study them—then he found out what he was up against, switched to paleobotany, and is now a curator of fossil plants at the Smithsonian.

Next in size to the brontotheres are several groups of rhinoceroses from the Badlands. There was the hippolike amynodont rhino *Metamynodon planifrons*, which was squat and short-legged like a hippo, but was a rhino relative nonetheless (figure 2.3*A*). The running hyracodont rhino *Hyracodon nebraskensis* was about the size of a Great Dane, with long slender legs for fast locomotion (figure 2.8*B*). In 1996, I published my research on these rhinos (Prothero 1996b) and found that far too many species had been recognized and given individual names. After measuring and plotting hundreds of skulls and jaws, I concluded that there were at most only five valid species of *Hyracodon*. Many of the so-called named species were based on inadequate specimens or on the fact that the crests of their upper premolar teeth are highly variable within a single population. Such variations cannot be used to distinguish species, however. Finally, there were rhinos that are members of the living family Rhinocerotidae, including the tiny *Penetrigonias*, the cow-size *Trigonias* (see figure 2.3*B*), the buffalo-size *Subhyracodon* (see figure 2.3*B*), and the even bigger *Amphicaenopus*. *Trigonias* is particularly

FIGURE 2.3 Hornless rhinoceroses of the Big Badlands: (*A*) *Metamynodon planifrons*, a hippolike form with big tusks and a semiaquatic lifestyle; (*B*) on the left, *Trigonias osborni*, a member of the living family Rhinocerotidae, with a full complement of incisors and canines in the front of its mouth (nearly all these teeth are absent in later rhinos); on the right, *Subhyracodon occidentalis*, the true rhinoceros that gave rise to the later lineages of living rhinos. (Drawings by C. Buell)

interesting because it still retains many of its primitive incisors and premolars in the front of the snout, teeth that are lost or modified in later rhinos. In the plains of northeastern Colorado are huge bone beds full of *Trigonias*, where museums have recovered literally hundreds of skulls and skeletons (in particular the Denver Museum of Nature and Science, the University of Colorado Museum in Boulder, and the University of California Museum of Paleontology at Berkeley). Russ Graham, formerly of the Denver Museum, has recently reopened these quarries with his famous "Bones Galore" excavations, and many more new specimens have come to light.

My particular favorite is *Subhyracodon* (figure 2.3*B*), the subject of my first research project, undertaken in 1976. (This animal is often mislabeled "*Caenopus*" in the older books and literature, but that name is invalid.) The name "*Subhyracodon*" is a bit of a misnomer—it is not related to hyracodonts, but is a true rhinocerotid. However, the rules of zoological nomenclature are very strict, and the inappropriate name given in 1878 cannot be thrown out just because it is confusing. Like *Trigonias* and *Hyracodon*, *Subhyracodon* was split into far too many species, again based on the highly variable upper premolar teeth. I first began examining these fossils at the beginning of my graduate career in 1976 and finally got my chance to straighten out this mess in my 2005 rhino book (Prothero 2005b). Now the number of valid species of this beast has been reduced to three (a fourth has been moved to its descendant genus, *Diceratherium*). None of these early rhinos had horns, despite our modern association of rhinos with horns among the five living species. Although rhino horns are made of compacted hairs and not bone (so they rarely fossilize), the horns leave a roughened surface on the top of the skull where they were attached, which gives us some indication of the horn's size and attachment.

Besides brontotheres and many different kinds of hornless rhinos, there were horses in the Badlands as well (figure 2.4). They are best known from *Mesohippus* (middle horse) and *Miohippus* (lesser horse) fossils. We've all heard the stories about horses evolving from beagle-size creatures of the early Eocene with four toes on their front feet and three on their hind feet. These creatures were formerly called "*Eohippus*" or "*Hyracotherium*," but are now known as *Protorohippus* and other names, according to David Froehlich (2002). *Mesohippus* and *Miohippus* are considerably larger (German shepherd size), with longer legs and especially reduced side toes, so that the central toe is much longer and stronger. Their skulls are noticeably more horselike in shape, with an elongate snout. Many of the teeth behind the front nipping incisors are missing, forming the gap where a horse takes its bit in its mouth. However, their teeth are only slightly larger than those of its primitive ancestors and have not yet developed the extremely high crowns that later horses would use to eat gritty grasses.

Back in the mid-1980s, my friend and former classmate Neil Shubin (now chair of the Department of Organismal Biology and Anatomy at the University

FIGURE 2.4 In the foreground, the gazelle-shaped humpless camel *Poebrotherium wilsoni*, a common denizen of the White River Badlands. In the background, *Mesohippus bairdi*, the collie-size, three-toed horse of the Big Badlands. (Drawings by C. Buell)

of Chicago) and I decided to tackle the problem of Oligocene horse evolution, using the amazing collection of complete skulls, jaws, and many skeletons in the Frick Collection of the American Museum of Natural History. We measured hundreds of specimens and coded the details of their teeth. When we finally had seen enough specimens, we began to realize that there were actually clear-cut distinctions between just a few valid species of *Mesohippus* that could easily be recognized from the upper molars and details of the skull. In 1989, we published our results, getting rid of many invalid species. More important, from the excellent Frick specimens we could tell that certain specimens had the special features of the ankle that distinguish *Mesohippus* from *Miohippus*. The two genera did not grade into one another, but were distinct lineages that overlapped in time. In one locality from the upper Eocene of Wyoming, we found evidence that three different species of *Mesohippus* and two of *Miohippus* had coexisted. As we'll see in chapter 4, the discovery of multiple lineages of horses that were static, unchanging, and overlapping in time and space strongly supports the punctuated equilibrium model of evolution, thus replacing an icon of gradual horse evolution with a totally different scenario.

Brontotheres, rhinos, and horses (plus primitive tapirs, which were rare at this time) made up the major lineages of odd-toed hoofed mammals, or perissodactyls. But far more common were the members of the order Artiodactyla, or even-toed hoofed mammals. The most abundant Badlands fossils by far are a group of extinct mammals known as oreodonts (figure 2.5). The lower Oligocene beds of the Badlands were originally known as the "turtle-oreodon" beds

FIGURE 2.5 The sheep-size, even-toed hoofed mammals known as oreodonts were the most abundant mammals of the Big Badlands. The most common one, *Merycoidodon culbertsoni*, is pictured here. (Drawing by C. Buell)

because of the great abundance of both of these fossils. Because oreodonts are extinct, it's tough to describe them to a nonpaleontologist. They were the size of a sheep, but distantly related to pigs or camels. They had teeth that looked like those of nothing alive today, short limbs, and feet with four splayed toes, not the two-toed hooves of ruminants.

Due to their great abundance, oreodont fossils are found in every rock shop around the world, and literally thousands of them are in the major museum collections of North America. For the longest time, very little could be done with oreodont fossils because they had been studied and classified by two paleontologists whose work sparked considerable controversy: C. Bertrand Schultz, long a professor and curator at the University of Nebraska, and Charles H. Falkenbach, a collector for the Frick lab. Schultz and Falkenbach published a series of detailed papers on oreodonts from 1940 to 1968 in the prestigious *Bulletin of the American Museum of Natural History*. As every paleontologist for the past 40 years who has analyzed their work has recognized (Harksen and Macdonald 1969; Lander 1977; Savage and Russell 1983; Gustavson 1986; Emry, Bjork, and Russell 1987; CoBabe 1996; Stevens and Stevens 1996; Prothero and Sanchez 2008), Schultz and Falkenbach typically did not adequately account for variation within a population when they created yet another questionable species or genus. In some cases, they did not account for the shape differences due to crushing of the skull after burial. Thus, they called specimens of the little oreodont *Miniochoerus* that were pancaked top to bottom *Platyochoerus* ("flat pig" in Greek) and named those that were lying on their side and crushed sideways

Stenopsochoerus ("narrow pig"). Fortunately, Margaret Stevens (now retired from Lamar University) spent years during the 1970s redoing Schultz and Falkenbach's work using modern statistics and population thinking (Stevens and Stevens 1996). She even made a soft latex cast of an uncrushed *Miniochoerus* skull and showed, by squeezing it in various directions, that Schultz and Falkenbach's taxa based on crushing could be explained by deformation. Now there is an updated taxonomy for *Miniochoerus* and *Merycoidodon*, the common Badlands oreodonts, as well as for many others, thanks to the work of these later paleontologists.

The oreodont family tree has many peculiar side branches. One group, the agriochoeres, was the most primitive of oreodonts, with relatively long robust limbs, a long tail, and claws rather than hooves on their feet. Some were the size of a chimpanzee, and many scientists have interpreted their peculiar anatomy as evidence that they climbed trees. My former student Joshua Ludtke began to tackle this project as part of his graduate work, and now he's completing a revision of this poorly understood group. The leptauchenines were smaller oreodonts about the size of a beagle, but with very peculiar heads. They had eyes high on the top of the skull, big openings for the nasal region, ears high on the back of the skull, and extraordinarily large bony chambers surrounding the middle ear. They also had the highest-crowned teeth of any mammal of that time except camels, suggesting that they ate a very gritty diet, and many of the specimens have teeth worn down completely to the roots. The peculiar shape of the skull—with the elevated eyes, ears, and expanded nasal opening—suggested to some paleontologists that leptauchenines were tapirlike or hippolike, spending their time immersed in water and bearing a short proboscis. However, the detailed evidence of the teeth, the enlarged ear chamber, and especially their occurrence in ancient dune deposits disproves the aquatic model and argues that they were probably desert dune dwellers instead. My former students Jonathan Hoffman and Francisco Sanchez and I recently completed a revision of this group (Hoffman and Prothero 2004; Prothero and Sanchez 2008).

Next in abundance after oreodonts in the Badlands were the tiny, hornless, deerlike ruminants known as *Leptomeryx* and *Hypertragulus*. These elegant creatures remind one of the modern "mouse deer" (genus *Tragulus*) or of the musk deer (genus *Moschus*), with their tiny size, long delicate limbs and toes, and lack of horns or antlers. Their minuscule jaws are incredibly abundant in the Badlands, although the skulls are much more rare than those of oreodonts. When Tim Heaton was a postdoctoral scholar at the Smithsonian, he undertook a major project deciphering the evolution of *Leptomeryx* (Heaton and Emry 1996). Tim found that only a few species—like other Badlands mammals that were common and therefore had been oversplit—were really valid and could be distinguished. The most common is *Leptomeryx evansi*, which Leidy named after John Evans, who had visited the Badlands in the 1840s.

After oreodonts and leptomerycids, the most common artiodactyl fossils in the Badlands are of the camels. These camels have the distinction of being first recognized and described by Leidy in 1847, the oldest of the Badlands mammals he named. But *Poebrotherium* and *Paratylopus* would not remind anyone of the living camel (see figure 2.4). They were small in size (comparable to a small gazelle) and, most important, did not have humps.

In fact, most ancient camels in general did not have humps. It's startling to think about rhinos without horns, camels without humps, and tiny three-toed horses evolving primarily in North America, but that's what the fossil record demonstrates. But the two living humped camels (the two-humped Bactrian camel of Asia and the one-humped dromedary of Africa) are atypical of camel history. More typical are the four types of camels found living today in South America: the llama, alpaca, vicuña, and guanaco. These camels are more lightly built, are faster runners, and lack humps. The specialized humps, feet, and other features of the two species of Old World camels are desert adaptations and not typical of the family as a whole.

Shortly after I finished my dissertation on the fossils and stratigraphy of the Badlands in 1982, I plunged into a study of the Badlands camels. The American Museum was full of beautiful skulls, jaws, and even complete skeletons of *Poebrotherium* and *Paratylopus* that had never been studied before, so it was paradise to work on the project. In fact, there's a whole floor of largely unstudied fossil camels in that collection, which I hope to be able to study and publish someday. It's one of the last great overdue revisions in all of mammalian paleontology. I soon realized that some species called *Poebrotherium* were actually *Paratylopus* and that many camels had been mislabeled "*Protomeryx*" based on a crummy specimen that couldn't be diagnosed. I worked my way up into a lineage of camels that included the long-misunderstood *Pseudolabis* and the amazing gazelle-camel *Stenomylus*. This latter creature had incredibly long delicate limbs and unbelievably high-crowned teeth whose roots reached to the bottom of the jaw and the top of the skull. I even had the pleasure and honor of naming an unusually large new species of stenomyline, *Miotylopus taylori*, after Beryl Taylor, the scientific staffer at the American Museum who had curated the entire camel floor. I had to cut the project off there so it could be published (Prothero 1996a), but there's a great deal of work still to be done because camels are among the most common fossils in the Oligocene and Miocene all over North America. Yet with the poor state of their taxonomy, most camel fossils cannot be correctly identified, and a huge amount of information is still unavailable.

The even-toed hoofed mammals included not only ruminating creatures with highly specialized teeth, but also many more primitive forms with simple low-crowned, blunt-cusped teeth. There were no true pigs (family Suidae) in the New World until the Spaniards brought them in the sixteenth century, but the New

World does have an ecological equivalent: the piglike creatures known as peccaries or javelinas (family Tayassuidae). Today, the three living species of peccaries reside mostly in Latin America, although they are also found in the desert Southwest parts of Arizona, New Mexico, and Texas. Like pigs, they are omnivores that eat plants, roots, seeds, fruits, and even meat or carrion when it's available, and, like pigs, they have simple teeth with low, rounded cusps for grinding and crushing a variety of food types. Peccaries and pigs had a common origin in Asia in the middle Eocene, and by the time of the late Eocene (the lowest stratum of the Badlands) peccaries had immigrated to North America, where they occupied the niche of a piglike omnivore (figure 2.6*A*). The earliest peccaries are my current project. They are found in the Badlands (*Perchoerus*) and in the John Day beds of Oregon (*Thinohyus*); they were the size of a small modern domestic pig, like the potbellied pig, and considerably smaller and more primitive than any living peccary. However, like the living species, the earliest peccaries had the enlarged upper and lower tusks that pointed straight up and down rather than the tusks of true pigs, which flare out from the side of the snout.

The peculiar extinct creatures known as anthracotheres form another piglike group (figure 2.6*B*). These animals vaguely resembled pigs and peccaries in their stocky low-slung bodies, long skulls, and broadly flared four-toed hands and feet. However, their teeth were a strange mixture of piglike low-crowned cusps with two W-shaped crests more like that of a camel or ruminant. They were relatively rare animals (only a few complete skulls of these creatures are known), yet they were ubiquitous in the Eocene and Oligocene of North America, Eurasia, and Africa. *Bothriodon* from the Badlands also occurred in Europe, and *Elomeryx* is found in Europe, Asia, and North America. North America also had some strange forms such as the long-snouted *Aepinacodon* and *Arretotherium*, and the peculiar creature from the Pine Ridge Reservation to which Reid Macdonald gave the name *Kukusepasutanka* (a Lakota mouthful that means "big nose hog").

Most interesting of all on this topic is the argument about what critters anthracotheres were related to. For a long time, paleontologists thought they were closely related to pigs, peccaries, and hippos, but no one could convincingly show which group was closer to them. Recent evidence, however, places them at the base of the hippo radiation, and Jean-Renaud Boisserie (2005, 2007) has shown that hippos were probably just specialized anthracotheres. This finding is consistent with these beasts' paleoecology. Most of their fossils are known to occur in river-channel sands or in swamp coals (the name Cuvier gave the genus in 1822, "*Anthracotherium*," itself translates to "coal beast"). Many had hippolike skulls with the eyes up on periscopes above the skull roof, high nasal openings, and ears at the top of the skull as well. Although many may have lived on water plants, the very last anthracothere genus *Merycopotamus* is very hippolike, and its fossil teeth even show evidence of grazing (Lihoreau 2003).

FIGURE 2.6 Piglike hoofed mammals of the Big Badlands: (*A*) the early peccary *Perchoerus probus*; (*B*) the hippolike anthracothere *Bothriodon*; (*C*) the gigantic, wart-faced entelodont *Archaeotherium mortoni*. (Drawings by C. Buell)

Beside camels, oreodonts, and leptomerycids, the most spectacular Badlands artiodactyls are the huge piglike creatures known as entelodonts, typified by *Archaeotherium mortoni* (figure 2.6C), which was the size of a large boar, and the huge Miocene *Daeodon* (formerly *Dinohyus*), which was the size of a rhino. Although entelodonts look superficially piglike, they are not members of the living pig family Suidae, and there is still no clear evidence as to where they belong in the artiodactyl family tree. Nicknamed "big pigs" or "pigs from hell," entelodonts had a face that only a mother could love. Their skulls were covered by all sorts of bony knobs and protuberances. Their massive cheekbones flared out broadly in many specimens, and some specimens had long flanges that stuck out from the side of the skull. What all these bumps and knobs were used for is anyone's guess. In some modern pigs and hippos, the knobs and bumps on their skulls are used for sparring with other males of their group, as well as to threaten and display their maturity and dominance in the social structure. Some entelodont skulls have deep punctures and long-healed gashes almost two centimeters deep that did not kill the animal, so these beasts definitely fought viciously with their big tusks. Their gapes were remarkable, capable of engulfing a rival's head, so some scientists think the large bumps and flanges on the skull protected it from being crushed and made it harder to be attacked, comparable to the way the flanges on a warthog skull protect its delicate areas around the eyes from damage. Some of these large flanges of bone in an entelodont skull, however, may have just been the insertion point for their huge powerful jaw muscles.

Entelodont teeth are also remarkable. They include huge canine tusks and large thick-enameled cheek teeth with blunt, rounded cusps. In mature entelodonts, the teeth have undergone tremendous wear, and the enamel is often broken. Even the huge tusks are worn down on the tips, so they were not kept sharp like the tusks in most mammals. Taken together, these teeth suggest a very rough diet, primarily crushed bones and scavenged meat. In many Badlands bone beds, there are bones that show the crushing effects of entelodont teeth. Their diet possibly included seeds, fruits, eggs, and whatever else they could find. Wear marks on some teeth suggest that they also chewed vines and roots for their water during the dry season.

Entelodonts may have been hoofed mammals that acted as scavengers and even part-time predators, but there were plenty of true predators during the Oligocene as well. One of the most remarkable was the wolf-size *Hyaenodon* (figure 2.7). As its name suggests, it looked superficially like a hyena with its heavy-limbed, low-slung build, powerful bone-crushing jaws, and long skull. However, *Hyaenodon* is not a member of the living order Carnivora (dogs, cats, bears, raccoons, weasels, hyenas, and their kin), but a member of a more primitive extinct order of carnivorous mammals known as creodonts. Indeed, it was the last of its line of creodonts because most of these archaic predators were in their hey-

FIGURE 2.7 *Hyaenodon horridus*, a wolf-size predator from the archaic extinct group of carnivorous mammals known as creodonts. (Drawing by C. Buell)

day during the Paleocene and early Eocene, and had nearly vanished by the end of the middle Eocene. The huge shearing teeth in its jaw were ideal for slicing flesh as well as crushing bones and were oriented so that they wore a continuous edge through the life of the animal, staying self-sharpening. In a number of places, concentrations of prey bones and fossilized feces with prey items such as oreodonts in them are believed to be the hyaenodonts' kill caches. Although hyaenodonts were impressive predators, they were clearly the last of their lineage and vanished from North America by the late Oligocene. They did, however, persist in Africa until the late Miocene.

True carnivorans (members of the living order Carnivora) were also abundant in the Badlands. The most common were the early members of the dog family Canidae, which included several different genera. You would never guess that the earliest dogs were an ancestor of Fido and Bowser, however. Dogs such as *Hesperocyon* (figure 2.8A) looked more like a weasel in their size and body proportions, with long, slender limbs and a very long, catlike tail. Only their distinctive teeth and skull features reveal the fact that they are primitive members of the dog family. There were no true cats yet, but there was a remarkable group of "false cats," or nimravids, which converged completely on the cat body form. Some, such as *Hoplophoneus* (figure 2.8D), was even saber toothed, although it was the size of a small leopard. Others, such as *Dinictis*, had shorter, straighter, more conical stabbing canines, not the curved blades of saber-toothed nimravids such as *Hoplophoneus*. For decades, paleontologists looked at these remarkable animals and concluded that they had to be cats. In the 1970s and 1980s, however, detailed studies were conducted of their skulls and especially the intricacies of their ear

FIGURE 2.8 True carnivorans of the Big Badlands: (*A*) *Hesperocyon gregarius*, an early representative of the dog family that was built more like a weasel; (*B*) the saber-toothed nimravid, or "false cat," *Hoplophoneus bellus*, shown here attacking the Great Dane–size running rhino *Hyracodon nebraskensis*. (Drawings by C. Buell)

regions, and it became clear that nimravids were not even remotely related to cats. Some studies even placed them on the dog branch of the Carnivora, along with bears, raccoons, and weasels! Their zoological affinities are still unsettled, but whatever they were related to, their catlike appearance is an amazing example of convergent evolution, when two different groups produce nearly identical anatomical features. Nimravids began declining in the late Oligocene, then vanished, so that for several million years of the early Miocene, there was a "cat gap," with no catlike predators known until true cats arrived in North America in the middle Miocene.

The array of large hoofed mammals and their predators in the Badlands is amazing, but numerically most abundant were the tiny mammals, especially the rodents and rabbits. They ranged from rabbits such as *Palaeolagus*, very similar to a modern jackrabbit, to a wide array of different rodents, including the squir-

FIGURE 2.9 Diorama of life during the Oligocene in the Big Badlands of South Dakota. (Painting by J. Matternes, courtesy Smithsonian Institution)

rel-like *Ischyromys*, and members of the true squirrel family Sciuridae to early relatives of the beavers, the sewellels, the gophers, the pocket mice and pocket gophers, and the hamsters and dormice. There were small insectivorous mammals as well, including primitive relatives of the hedgehogs, moles, shrews, and several extinct groups. These fossils may not be as big or spectacular as the huge brontotheres and saber-toothed nimravids, but ecologically they were extremely abundant and important.

So what did the world of these Badlands creatures look like? Most reconstructions (figure 2.9) are based on evidence from the ancient soils and some plant fossils, which indicate a mixture of thick forests along riverbanks and more open scrublands between the trees, but no true open savannas or grasslands. In some places, the sedimentary evidence suggests that there were wind-blown dunes, especially in the later Oligocene. Volcanic ash falls from Utah and Nevada frequently blanketed the ground, and there are numerous distinctive volcanic ash layers throughout the sequence. Most of the animals were apparently buried on floodplains right after flooding events, although the stream dwellers are sometimes fossilized in the sandstones of ancient river channels, where they were washed in and buried.

A number of studies have been done to reconstruct the animals' ancient ecology as well. During the late Eocene, the region was wet and forested enough to supported occasional crocodiles as well as numerous pond turtles. The obligate leaf-eaters, or browsers, included the giant brontotheres, the tapirs, the rhino

Amphicaenopus, and probably the anthracotheres; all are found primarily in the river-channel sandstones. By contrast, the majority of the Badlands mammals, especially the oreodonts, camels, ruminants, horses, hyracodonts, and their predators apparently lived in more brushy, scrubby open country. By the early Oligocene, the brontotheres had vanished, so the riverside forests were inhabited mostly by tapirs, anthracotheres, and the rhino *Metamynodon*. Some animals, such as the true rhino *Subhyracodon*, may have lived in both the forests and the scrublands.

These paleoecological inferences are based largely on the animals' body shape and size and by details of their teeth. In the past few years, however, paleoecology has been revolutionized by a wide variety of new techniques that shed new light on how ancient animals lived and what they ate. These techniques range from studies of the sharpness and wear on the cusps (mesowear) to microscopic views of the pits and scratches on the tooth enamel caused by different diets and feeding habits (microwear) to analysis of the detailed chemistry of the tooth enamel itself to see what plants the animal ate.

Among the more recent studies, Eric Dewar (2007) analyzed the microwear of a wide variety of Badlands mammals. He found that most of the carnivorous mammals (even those with teeth specialized for shearing such as the nimravids) show lots of scratches and pits from bone eating, so few were exclusively meat eaters, as most living carnivorans are today. Likewise, he found that lineages that did not change their overall tooth shape over time did show signs of changes in diet nonetheless, with diets of pure browse (leaves) being gradually replaced by more gritty diets of shrubs and grasses. A number of studies of the oxygen and carbon isotopes trapped in these teeth (Zanazzi et al. 2007) have shown that nearly all the herbivores ate bushes, shrubs, and leaves; none of the teeth had the chemical signature of modern temperate and tropical grasslands. This avenue for future research is very promising and bound to give paleontologists answers to questions that have puzzled us for more than a century.

The Time Dimension in the Badlands

Reconstructing the paleoecology of Badlands mammals is an interesting and important thing to do, but there is an additional dimension that modern ecologists do not have the data to consider: deep time. As we shall see in chapter 3, the White River sequence ranges from 37 million years to about 30 million years in age, so considerable time passes as you move up layer by layer in these rocks. Early collectors quickly began to notice that the fossil assemblages at each level were different, so it was clear that they changed through time. By 1900, Henry Fairfield Osborn and his colleague at the American Museum, the brilliant paleontologist William Diller Matthew, had subdivided the Badlands sequence into a series of "Life Zones" (figure 2.10), which roughly corresponded to the different rock units: the Chadron Formation and the overlying Scenic and Poleslide members of the Brule Formation. The "*Titanotherium* zone" was the Chadron Formation (then considered early Oligocene, but now identified as late Eocene [chapter 3]), the "*Oreodon*" and "*Metamynodon* zone" was the lower Brule Formation, and the "*Protoceras*" and "*Leptauchenia*" was the upper Brule Formation.

Osborn and Matthew were following a tradition that had been common in paleontology (especially invertebrate paleontology) since the late 1790s. At that time, the English canal engineer William Smith was the first to observe that the fossil record changes through time (called *faunal succession*) and that each formation had its own distinctive and nonrepeating assemblage of fossils. The French paleontologists Baron Georges Cuvier and Alexandre Brongniart made the same discovery in the Paris Basin of France around 1805. Soon thereafter, enthusiastic, hard-working paleontologists all over Europe were documenting the sequence of fossil assemblages in each stack of formations and using the fossils to date the rocks and correlate them around the world (the practice known as *biostratigraphy*). In the process, they had abandoned the old, simplistic notions that the world's rocks had been formed during Noah's flood. In fact, some of them (such as Alcide d'Orbigny) had shown that at least 29 separate floods are not mentioned in Genesis and could not be squeezed into the biblical accounts. Ironically, most of these pioneering paleontologists in the 1820s through 1850s were relatively conventional, devout men, who had no incentive to question the Bible and were not out to disprove the Noah's flood story or provide evidence for evolution. Indeed, the Darwinian theory of evolution wasn't published until 1859, long after most geologists and paleontologists had completely abandoned the Noah's flood story as disproven by the rock record.

One of the most influential pioneers of biostratigraphy was the German geologist Albert Oppel. In the 1850s, he published his work on the Jurassic fossils of the French Alps near Geneva and developed a new method of biostratigraphy. Instead of compiling tables of fossils found in each formation, as Cuvier and d'Orbigny had done, he followed the lead of his mentor Friedrich Quenstedt and plotted the *range* (the total vertical stratigraphic distribution of a fossil species) of each species throughout many formations. Each species had its own range in time and space, and by plotting the overlapping distributions of many different ranges, Oppel was able to subdivide time in the Jurassic into many small units, or *range zones*. For example, the interval between, say, 10 and 20 meters (33 and 66 feet) on a given section might be characterized by the first appearance of fossil X and the last appearance of fossil Y (figure 2.11). That interval might be called

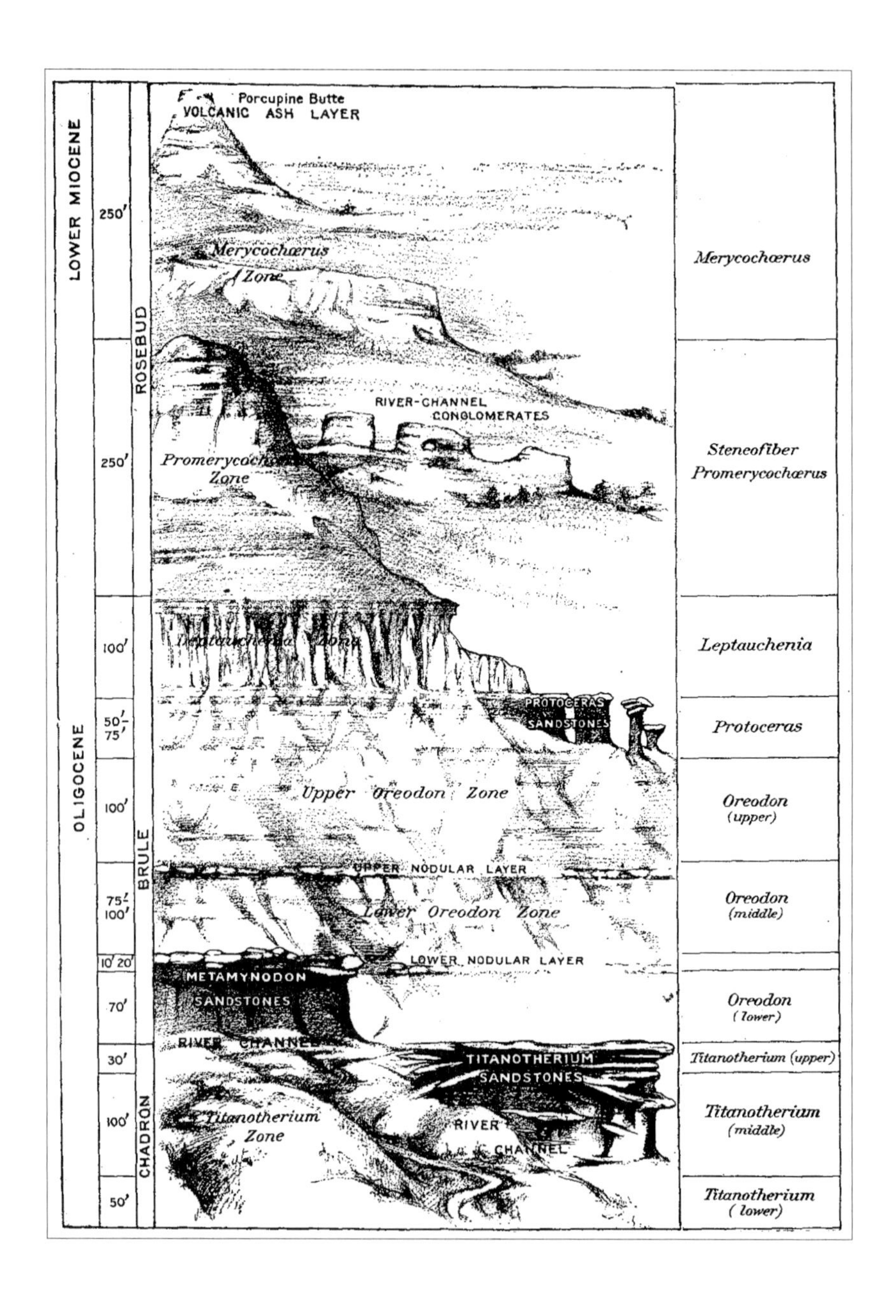

FIGURE 2.10 The "Life Zones" developed in 1899 by William D. Matthew and Henry Fairfield Osborn as a biostratigraphic scheme for the Big Badlands. (From Osborn 1929)

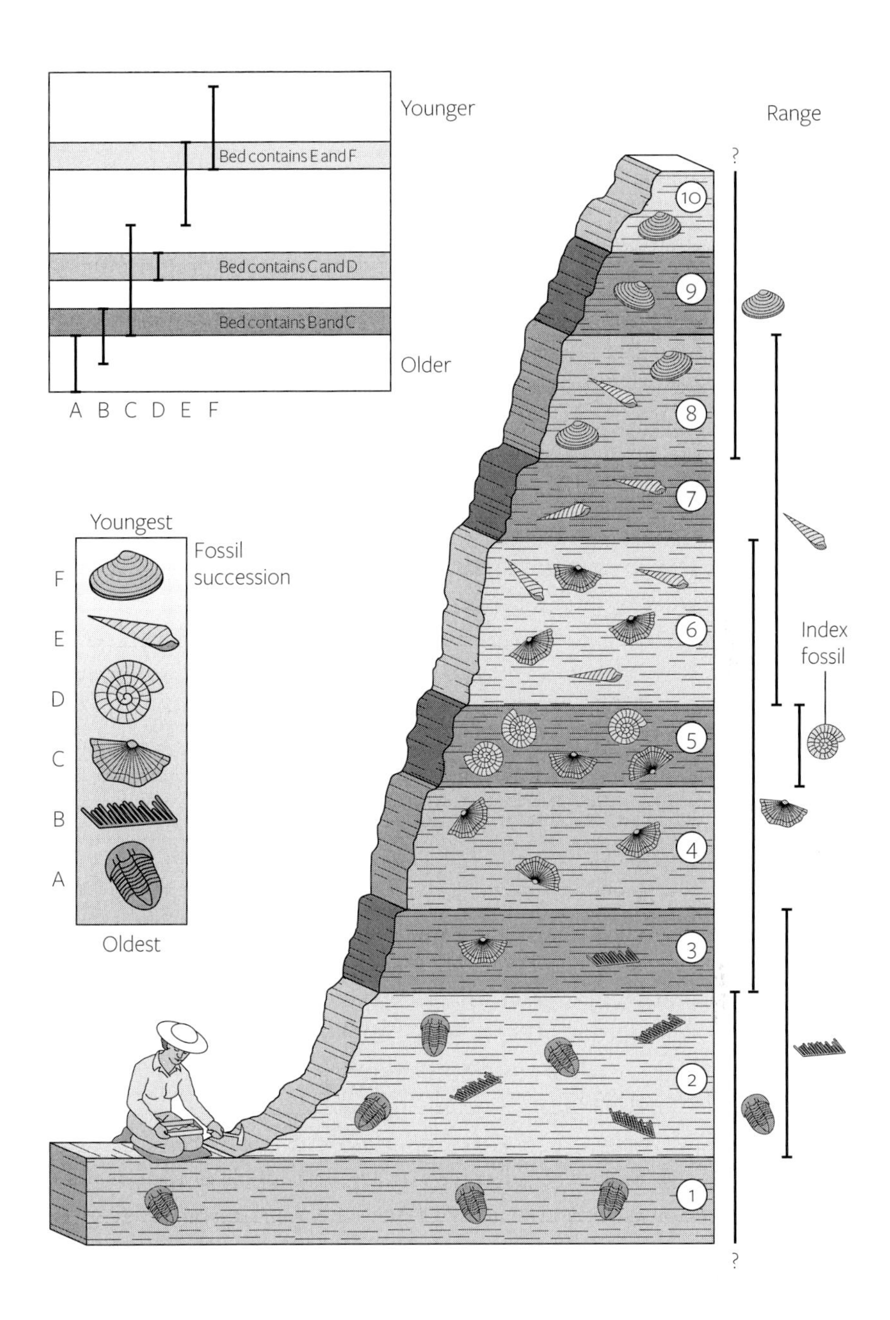

FIGURE 2.11 The overlapping range-zone method of biostratigraphy: (*a*) by plotting the vertical ranges of a series of fossils in each stratigraphic section, (*b*) their distribution through time can be documented. The overlap between the first appearance of one fossil and the last appearance of another defines a range zone. The more such range zones that can be defined in a given area, the finer the resolution and detail and the better we can subdivide geologic time.

the "X–Y range zone" in a standard scheme of defining range zones by their fossils. The more of these fine-scale subdivisions based on overlapping ranges of many different fossils you can define, the shorter each zone is and the better the time resolution of your biostratigraphy. Oppelian range-zone biostratigraphy has become the standard method by which most biostratigraphy (especially of invertebrate fossils) has been practiced ever since Oppel devised it. Indeed, it is bread and butter in the oil companies, which plot the biostratigraphy of hundreds of different microfossils to allow fine-scale correlation of the beds from one drill core to another. Indeed, without Oppel's technique, we would have no oil today!

However, vertebrate paleontologists seemed to march to the beat of a different drummer in this regard. The early paleontologists of the late nineteenth century, such as Leidy, Marsh, and Cope, relied on field parties to bring fossils to them and only rarely visited the field or collected their own specimens. None of the early collectors had much detailed information about where they were geographically, and they often didn't even know what formation the fossil came from, so the early specimens were useless for biostratigraphy. By 1900, however, enough crews from the American Museum and other leading institutions had worked out the general stratigraphy of the Badlands and other places that they could create a rough zonation of fossils in the western United States, although most of their "zones" were roughly equivalent to formations or members (see figure 2.10), and no finer subdivision could be attempted with such early collections and such poor field data.

Then the vertebrate paleontology community took a step back from Matthew and Osborn's promising start. By the late 1930s, a committee chaired by Horace E. Wood II formally established a mammalian timescale for North America. The final Wood Committee report (Wood et al. 1941) erected a series of "land mammal ages" that were not true Oppelian range zones or ages/stages in the traditional sense used by all other paleontologists, but an unfortunate mix of biostratigraphic units and rock units. For example, it defined the Chadronian land mammal "age" on the basis not only of the time when *Mesohippus* and brontotheres coexisted (a biostratigraphic definition), but also of the time span of the Chadron Formation (a rock stratigraphic definition). They assumed that these two sets of criteria would always remain the same, but using the double set of definitions was a disaster waiting to happen. Sure enough, in 1951 Morris Skinner found brontothere fossils in rocks lying above the Chadron Formation. So how was the Chadronian to be defined? Based on the last brontothere fossil or on the top of the Chadron Formation? These two events no longer coincided, so the Wood Committee definition was in trouble.

Not long after the "land mammal ages" were established, paleontologist Don Savage at Berkeley became unhappy with their implications. Berkeley at that time had the only collegiate Department of Paleontology in the country, and it was a hotbed of biostratigraphic ferment. Invertebrate paleontologists who practiced classical Oppelian biostratigraphy showed their vertebrate paleontologist colleagues the inadequacies of the Wood Committee's concepts. In 1955, Savage published an influential paper in which he attempted to name Oppelian range zones for some Miocene beds in California, and in 1962 he published (in an obscure Argentinian journal impossible to get in the United States) an impassioned argument for using range-zone biostratigraphy rather than "land mammal ages" in vertebrate paleontology as well.

His plea was influential on many of the Berkeley students of the 1950s and 1960s, who went on to become the next generation of museum curators and professors. My undergraduate professors in paleontology were Mike Woodburne and Mike Murphy at the University of California, Riverside, both of whom earned their Ph.D.s at Berkeley. In graduate school, my advisers were Malcolm McKenna and Richard Tedford, also Berkeley students in the 1950s and 1960s. Throughout my college education, I heard again and again about the importance of range-zone biostratigraphy, the inadequacy of the Wood Committee methods, and the need to get the most finely resolved, precisely dated sequence of fossils that our data would allow.

That quest has inspired many of us who were either trained at Berkeley or were students of Berkeley graduates (as I am). For example, in the 1960s John Rensberger developed in his Berkeley dissertation a formal range-zone biostratigraphy of the famous late Oligocene–Miocene John Day beds of Oregon. Everett Lindsay proposed in his Berkeley dissertation a biostratigraphic zonation for the middle Miocene beds near Barstow, California. Other proposals for local range-zone biostratigraphies in North America were suggested as well.

Unfortunately, there was a good reason why the Wood Committee didn't adopt a fully Oppelian system back in 1941. Unlike most invertebrate fossils, such as trilobites and mollusks or microfossils that tend to be abundant and easy to collect through many levels of a given formation, vertebrate fossils are usually much more rare and not distributed uniformly through a formation. Instead, they tend to occur in local quarries or bone beds or other concentrations, and there are many feet of strata above and below them that are barren of vertebrate fossils. In such cases, you can't finely subdivide the rock section with fossils because the fossils occur at only a few levels. In many other cases, however, the problem was simple laziness or lack of precision. For more than a century, people would pick up fossils without recording the exact stratigraphic level from which each fossil came. Once they left the region, it was impossible to reconstruct each fossil's exact stratigraphic position, and range-zone biostratigraphy was impossible.

Such was the situation in the 1940s when Morris Skinner (figure 2.12) began collecting fossils in the White River Group. Skinner was a rambunctious

FIGURE 2.12 Morris Skinner in the Badlands around 1950. (Photograph courtesy M. Skinner)

teenage farm boy from Ainsworth, Nebraska, when he found a mastodon and soon became interested in collecting fossils. By the 1930s, he had been hired by the millionaire Childs Frick, son of the railroad baron Henry Clay Frick, Andrew Carnegie's partner, to collect fossils all over the western United States. Childs Frick, like many rich kids at that time, had started out as a big-game hunter, then became interested in the origins of the animals he was hunting and eventually obtained a degree in paleontology at Berkeley in the 1920s. Frick's money kept field crews going all year round in the American West. In the summers, Skinner, along with a bunch of other former midwestern farm boys such as Ted Galusha, Charlie Falkenbach, and Joe Rak, would undertake huge explorations in the northern Plains, especially in western South Dakota, Nebraska, and eastern Wyoming. In the winters, they would escape the snow and work in New Mexico, Arizona, California, and Texas.

Frick's crews amassed an incredible collection of fossils, which were part of Frick's private collection until he died in 1965. Then his endowment helped build the Frick Wing of the American Museum (where he had long been a trustee), which contains ten floors for paleontology research. The upper three floors

FIGURE 2.13 Typical drawer (one of dozens of such drawers) of oreodont skulls (*Miniochoerus gracilis*) in the Frick Collection of the American Museum of Natural History. Painted in the patch of white on every specimen is the exact stratigraphic data for that specimen, a level of detail not found in any other collection. (Photograph by the author)

are offices, libraries, classrooms, and a preparation lab for cleaning and restoring fossils. Seven floors are storage of fossil mammals—a floor each for mastodonts and mammoths, rhinos, horses, camels, and artiodactyls other than camels, and two more floors with all the rest of the mammals. Even more impressive than the quantity of the fossils is the quality. Where we used to have a few isolated teeth or jaws of a given species, now we typically have many skulls, jaws, and even complete skeletons (figure 2.13). Everything we think we know about North American middle and late Cenozoic mammals is actually out of date until the Frick Collection is fully studied. Some groups have received a great deal of interest (e.g., horses), and I spent 20 years documenting and revising the North American rhinos (Prothero 2005b). Other groups, such as the camels and pronghorns, are badly in need of study, but that task will take many more years in New York to finish.

Among Frick's collectors, Morris Skinner was extraordinary. Not only did he have a gift for finding fossils and a talent for managing huge excavations and recovering beautiful complete specimens where only scraps were known previously (especially in western Nebraska), but he was also a first-rate stratigrapher. As soon as he began collecting an area, he measured the thickness of the local stratigraphic section and recorded the details in his notebook. He and his crews would then learn the stratigraphy and be sure to record the exact level from which each fossil had come, usually to the nearest foot above or below a marker bed. One of his crew was another former farm boy from Ainsworth, Nebraska, Bob Emry. Bob and his brother Raleigh started by collecting fossils with Morris when they were just teenagers and soon became regular members of his crew. With Morris's encouragement, Bob went on to finish his bachelor's degree at Colorado State and then came to Columbia and the American Museum in the 1960s, where he did his dissertation on one of Morris's favorite localities, Flagstaff Rim, near Casper, Wyoming. Bob then became a curator at the Smithsonian, and he was my mentor when I was his field assistant in the White River beds in 1978.

Morris was always a stickler for being careful about stratigraphic data collected in the field. After generations of sloppy work by vertebrate paleontologists, he knew that it was essential to record the precise data on every specimen at the time of collection. One of the stories Bob told me was about one of his first finds in the White River beds near Lusk, Wyoming. An eager young teenager, Bob showed Morris a nice oreodont skull he had found. When Morris asked him if he knew precisely what level the specimen had come from, and Bob admitted that he hadn't noticed, Morris threw the fossil over his shoulder and pronounced it worthless. With harsh lessons like this, all of Morris's crews were careful biostratigraphers from early on in their careers.

Morris started working the Big Badlands of South Dakota in the 1940s, when few other groups worked out there (especially during the World War II years). He and his crews built up enormous collections there, and he worked out the detailed stratigraphy of many parts of the Big Badlands that had never been studied previously. By the 1950s, he was working mostly in Wyoming and Nebraska, where the fossil beds are even richer and more complete than those in South Dakota. Many of these localities had never been collected, so they were rich in specimens (although this is no longer true). Many of these beds were on private ranch land, and Morris, with his "aw shucks" farm boy manners, could easily charm the ranchers into cooperating with him, although they might not have been so friendly to arrogant college-educated scientists out of the big universities in the East and the West.

By the time Frick died in 1965, the Frick Collection from the White River beds was immense, yet almost completely unstudied. It took most of a decade to raise the money and build the Frick Wing, then curate all the fossils within it. When I arrived at the American Museum in 1976, the Frick Collection was finally available for study to anyone who had years to spend doing research in New York City.

I first started working on Jurassic mammals and lizards for my master's project, but as I began to look around for a suitable Ph.D. project, the riches of the Frick Collection beckoned. I could see that the White River fossil collection was enormous, that it had excellent stratigraphic data on every specimen, and that no one had ever studied it (except for Schultz and Falkenbach with their awful oreodont work). With the new technique of magnetic stratigraphy (chapter 3), I felt that a good dissertation project would be a biostratigraphic zonation of the White River Group and improvement of its dating. I spent the field summer of 1978 with Bob Emry learning all I could about the Frick White River localities, and by the next summer I was doing my own research.

I didn't realize it at the time, but the White River fauna was overdue for this kind of research. As noted earlier in the chapter, the late nineteenth and early twentieth centuries were the heyday of Badlands research, when many paleontologists cut their teeth on Badlands mammals. This fieldwork culminated in a series of monographs (five volumes published between 1937 and 1941) by William Berryman Scott, Glenn Lowell Jepsen, and Albert E. Wood. After that, the White River beds became a battleground. A number of groups worked in the area, but they were at odds with one another and constantly sniping at each other's research. A contentious figure such as C. Bertrand Schultz was constantly at odds with other paleontologists, including the strong-willed J. Reid Macdonald at the South Dakota School of Mines in Rapid City and the cantankerous John Clark, who spent his career at the Carnegie Museum in Pittsburgh, the Field Museum in Chicago, the University of Colorado Museum, and many other institutions. Finally, there were Morris Skinner and his colleagues at the Frick lab, who mostly worked quietly and kept to themselves, never publishing much (because Frick forbade it), but doing the best scientific work and collecting of any of these warring parties.

By the 1960s, the political and scientific warfare among Schultz, Macdonald, and Clark was so intense in print and behind the scenes that almost no student could work on White River fossils or rocks without getting caught in the crossfire. Consequently, research on these amazing fossils started to decline, especially as Schultz, Macdonald, and Clark reached retirement age, and no students were in the pipeline to replace them. By the time I began studying the White River in 1978, only Bob Emry was available to mentor me, and he had escaped being wounded in the White River wars by his years as a Frick collector.

Thus, I walked into an ideal situation: an immense collection of unstudied White River fossils. Nearly all of them had excellent stratigraphic data compared to every other White River collection in every other museum or rock shop, which was useless for doing range-zone biostratigraphy. By 1978, I was eagerly going through the collections, identifying the fossils, and plotting their biostratigraphic ranges (figure 2.14). By 1982, I compiled all this information in my dissertation. It took a number of years getting other scientists to finish revising the

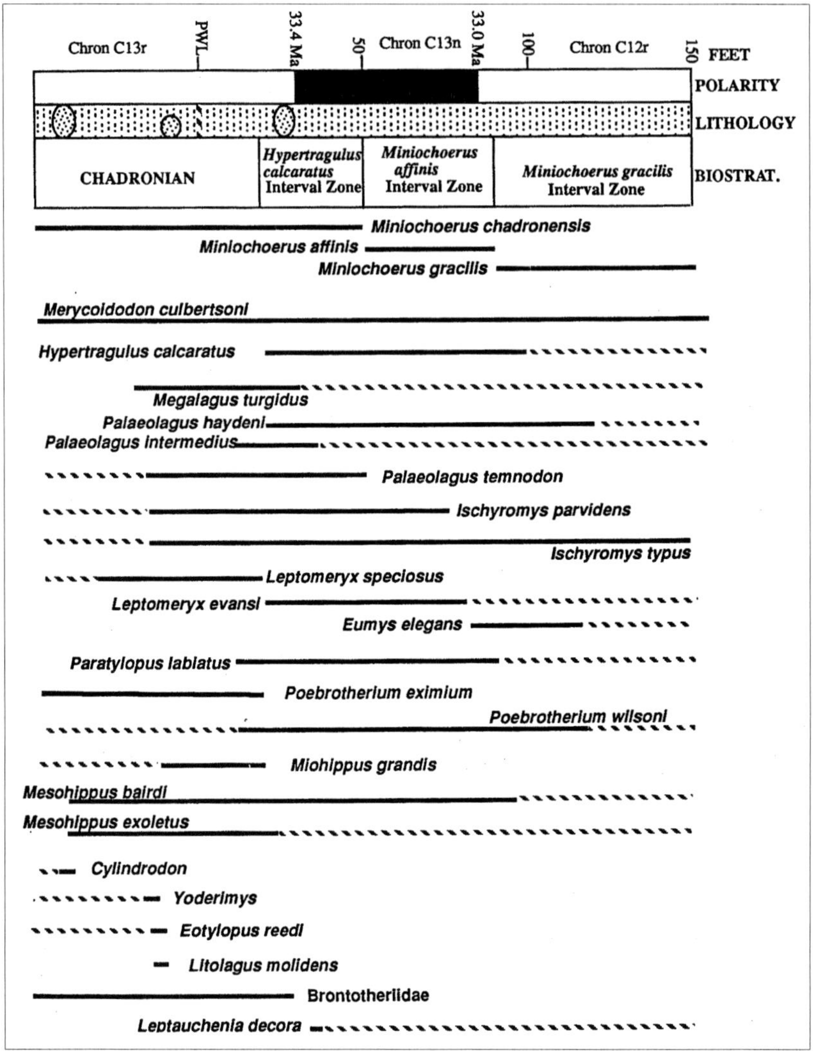

Chron C13r
PWL
33.4 Ma
50
Chron C13n
33.0 Ma
100
Chron C12r
150
FEET
POLARITY
LITHOLOGY
CHADRONIAN
Hypertragulus calcaratus Interval Zone
Miniochoerus affinis Interval Zone
Miniochoerus gracilis Interval Zone
BIOSTRAT.

Miniochoerus chadronensis
Miniochoerus affinis
Miniochoerus gracilis
Merycoidodon culbertsoni
Hypertragulus calcaratus
Megalagus turgidus
Palaeolagus haydeni
Palaeolagus intermedius
Palaeolagus temnodon
Ischyromys parvidens
Ischyromys typus
Leptomeryx speciosus
Leptomeryx evansi
Eumys elegans
Paratylopus labiatus
Poebrotherium eximium
Poebrotherium wilsoni
Miohippus grandis
Mesohippus bairdi
Mesohippus exoletus
Cylindrodon
Yoderimys
Eotylopus reedi
Litolagus molidens
Brontotheriidae
Leptauchenia decora

FIGURE 2.14 Range-zone biostratigraphy for one of the key White River sections in the Seaman Hills north of Lusk, Wyoming. (After Prothero and Whittlesey 1998:fig. 6)

taxonomy of the key groups of mammals (published in a volume I edited with Bob Emry in 1996), but Emry and I were eventually able to develop a completely Oppelian range zonation for the White River fauna (Prothero and Emry 2004; see also Prothero and Whittlesey 1998). Little did I know at the time, but the White River fauna would eventually become the key to many other important scientific problems. I had picked the project just because I knew that White River fossils and stratigraphy were understudied, and yet of fundamental importance, but you never know where fate will lead you.

I was extremely fortunate in starting my career when I did because all of the old guard was still alive and I met them all. I saw Schultz only at a distance at SVP meetings until he stopped showing up. Reid Macdonald was a curator at the Los Angeles County Museum of Natural History when I was an eager young kid dying for advice on a career in paleontology. I sent him a letter, and he replied with a highly detailed, hand-typed, three-page letter full of good advice that has helped me throughout my career. I later got to know him much better in his retirement at his Flying *Aepinacodon* Ranch outside Rapid City before he died in 2004. John Clark had dropped out of sight for years, but I unexpectedly came across him when I was visiting the Big Badlands in 1983. My crew and I were sitting in on a campfire talk by the rangers for the general public visiting the park. The old guy giving the campfire talk seemed to know stuff that no ranger would, and by the end I realized it was John Clark (I had missed his introduction). I introduced myself and got some advice on how to reach his classic Chadron sections west of Sheep Mountain Table. Then, when I was working in the Big Badlands in the summer of 1986 I ran into him again, and he spent several hours with my crew and me in the living quarters for Badlands employees, eagerly showing us maps of what he had been doing. This was the last time I would see him before he died in 1996 at age 80.

Of the old guard, I was closest of all to Morris Skinner, whom I met when I first arrived at the American Museum in the fall of 1976. I got to know him much better when I was Bob Emry's field assistant in Nebraska in the summer of 1978. Bob would take me over to Morris's house almost every time we returned to Ainsworth that summer, where Morris would regale me with stories about the days of the Frick lab. I saw him frequently after that because he spent most winters in New York getting his years of knowledge ready for publication. He published two huge papers on the stratigraphy, geology, and fossils of his most important collecting areas in western and central Nebraska (Skinner, Skinner, and Gooris 1977; Skinner and Johnson 1984), but none of his ideas about the White River reached print. Instead, they all were stored in his unpublished field notebooks in the American Museum archives; I consulted them often during my research and made copies of nearly every page of crucial localities and sections. When Morris passed away at age 83 in 1989, I felt as if I had walked in his footsteps for more than a decade. By then, he had been given an honorary doctorate by the University of Nebraska, and many fossils had been named after him. Shortly after his death, the SVP's award for excellence in fieldwork was named the Morris F. Skinner Award.

A coda. In 1994, during the talks at a big symposium hosted by Bob Hunt and Dennis Terry at the University of Nebraska at Lincoln on the White River and

Arikaree beds and their fossils, each author ended up praising the foresight and brilliance of Morris Skinner's largely unpublished work that other scientists had confirmed six years after he had died and emphatically rejecting the ideas of Schultz and his group. Schultz himself had died just months earlier. No matter how well connected and forceful a scientist may be, later investigators will eventually determine who had the better ideas and whose work will stand the test of time.

Further Reading

Emry, R. E. 2002. *Good Times in the Badlands*. San Jose, Calif.: Writer's Showcase.

Galusha, T. 1975. Childs Frick and the Frick Collection of Fossil Mammals. *Curator* 18, no. 1:5–38.

O'Harra, C. [1920] 1976. The White River Badlands. *South Dakota School of Mines Bulletin* 13:1–181.

Prothero, D. R. 1994. *The Eocene–Oligocene Transition: Paradise Lost*. New York: Columbia University Press.

Prothero, D. R. 1996a. Camelidae. In D. R. Prothero and R. J. Emry, eds., *The Terrestrial Eocene–Oligocene Transition in North America*, 591–633. Cambridge: Cambridge University Press.

Prothero, D. R. 1996b. Hyracodontidae. In D. R. Prothero and R. J. Emry, eds., *The Terrestrial Eocene–Oligocene Transition in North America*, 634–645. Cambridge: Cambridge University Press.

Prothero, D. R. 2005. *The Evolution of North American Rhinoceroses*. Cambridge: Cambridge University Press.

Prothero, D. R., and R. J. Emry, eds. 1996. *The Terrestrial Eocene–Oligocene Transition in North America*. Cambridge: Cambridge University Press.

Prothero, D. R., and R. J. Emry. 2004. The Chadronian, Orellan, and Whitneyan North American land mammal ages. In M. O. Woodburne, ed., *Late Cretaceous and Cenozoic Mammals of North America*, 156–168. New York: Columbia University Press.

Prothero, D. R., and K. E. Whittlesey. 1998. Magnetostratigraphy and biostratigraphy of the Orellan and Whitneyan land mammal "ages" in the White River Group. In D. O. Terry, H. E. LaGarry, and R. M. Hunt Jr., eds., *Depositional Environments, Lithostratigraphy, and Biostratigraphy of the White River and Arikaree Groups (Late Eocene to Early Miocene, North America)*, 39–61. Geological Society of America Special Paper no. 325. Boulder, Colo.: Geological Society of America.

Rainger, R. 2004. *An Agenda for Antiquity: Henry Fairfield Osborn and Vertebrate Paleontology at the American Museum of Natural History*. Tuscaloosa: University of Alabama Press.

Terry, D. O., H. E. LaGarry, and R. M. Hunt Jr., eds. 1998. *Depositional Environments, Lithostratigraphy, and Biostratigraphy of the White River and Arikaree Groups (Late Eocene to Early Miocene, North America)*. Geological Society of America Special Paper no. 325. Boulder, Colo.: Geological Society of America.

The crucial section at Flagstaff Rim, west of Casper, Wyoming, in 1977. The white bands across the outcrop are the volcanic ashes that provided the first potassium-argon dates and later the argon-argon dates for the Chadronian (late Eocene) in North America. The white band just below the dark vegetated area at the top is Ash J, which yields an argon-argon date of 34.479 ± 0.087 million years. Midway down the section is Ash G, which yields a date of 35.574 ± 0.027 million years. At the base of the section (in the wash below the horizon) is the oldest dated unit, Ash B, which yields a date of 35.973 ± 0.224 million years. As discussed in this chapter, these dates were crucial to recalibrating the North American timescale in 1989. In the foreground are Malcolm McKenna (*left*) and John Flynn (*right*). (Photograph by the author)

3 | Magnets and Lasers

Stratigraphy is the backbone of the science. Without sound stratigraphy, structural studies are impossible, and all historical and most economic geology depend upon it.

—JAMES GILLULY, "AMERICAN GEOLOGY SINCE 1910—A PERSONAL APPRAISAL"

The Curse of Crazy Johnson

June 25, 1983, dawned warm and clear, like so many other summer days in the Badlands. We were camped near the visitor's center just south of Cedar Pass, the headquarters of Badlands National Park. My field crew consisted of my good friend Annie Walton, an Amherst grad planning to enter grad school in paleontology; my former Vassar student Allison Kozak, who had been on my seven-week Vassar geology cross-country trip in the summer of 1981; and my current student Rob Lander, who was finishing his bachelor's degree in geology at Knox College in Galesburg, Illinois, where I was teaching at the time. I had just received a grant from the Petroleum Research Fund of the American Chemical Society to do my research, so my crew and I were headed all over the western United States to tackle as many projects as we could in a few weeks. Over the course of that field season, we had been all over the White River beds in Wyoming, Nebraska, and South Dakota, including a return trip to Flagstaff Rim, Wyoming to resample this crucial section. Then we went to the upper Oligocene John Day beds of central Oregon, where we celebrated a great small-town Fourth of July, complete with a parade and fireworks. Then we headed north to the Olympic Peninsula of Washington to collect in marine beds and finally back through Montana to sample more sections.

This trip was typical of my field operations since I had begun my own research in the summer of 1979. In my first field season, I was only a third-year grad student on a shoestring budget. I had hardly any funding, so I recruited two undergraduate friends of mine from Columbia University, Heidi Shlosar and Karen Gonzalez, who worked for just their paid expenses and the experience of camping and collecting out west. We had spent several weeks working Morris Skinner's main Frick localities along the Pine Ridge in eastern Wyoming and western Nebraska. At each locality, we were collecting lots of small samples of rock, spaced every few feet on the stratigraphic section as we climbed up the ravines and ridges. We scraped a horizontal surface on top of each rock and marked a north arrow on that surface.

During this first collecting trip, we had our share of close calls and hair-raising incidents. To save money, we would pitch our tents in the pastures of the rancher who had given us permission to work on his land. In one case, we discovered that the pasture had an angry bull in it, and we just barely made it over the fence before he charged. In other cases, we worked in rugged badlands far off the good roads, southeast of Douglas, Wyoming, before the place was subdivided into ranchettes. One day, we had just finished collecting a long section through the Reno Ranch–Wulff Ranch area, and I had left the outcrop well ahead of my crew to take the truck to town for errands. In that first field season, I was driving an old, beat-up 1966 Ford pickup that had once been driven by Morris Skinner for the Frick Lab, but was now stored at the American Museum in New York. My adviser had given me permission to use it because it had four-wheel drive (rare on vehicles then), even though it had a manual transmission and only six cylinders and well more than 100,000 miles on it. The Reno Ranch–Wulff Ranch sections were then in a deep valley with only a steep dirt road leading in and out. Getting in was no problem, but when I gunned the engine and tried to drive up that steep slope, I lost power about halfway up and suddenly found myself rolling backward down this steep narrow dirt road. The brakes were not strong enough to keep the truck under control, and I had no engine power to use my gears to slow down (not that I could do anything, or I'd strip the clutch). I did the best I could steering backward downhill and made it to the bottom, only to go off into the sand where the road made a sharp bend. There I was stuck, so I waited for my crew to arrive.

Once they caught up with me, we did all the traditional tricks for getting unstuck, and soon I was on the dirt road again. This time I took no chances. I gave myself a long approach run to the slope, so I was going uphill as fast as I could, and I made it all the way to the top without losing power. Then my crew got into the truck, and we continued on to our next adventure.

Three weeks of fieldwork ended, and we headed back to New York and the American Museum. There I took all the samples and trimmed them down into cubes with a tungsten-carbide bladed band saw. Once I had prepared all the samples, I took them to a magnetometer then located at Woods Hole Oceanographic Institute in Cape Cod (figure 3.1). My friend Chuck Denham had a paleomagnetics lab that measured the direction and intensity of the magnetic field trapped in

FIGURE 3.1 The old cryogenic magnetometer at the Woods Hole Oceanographic Institute as it appeared in 1979. (This lab is now dismantled.) The large cylinder is surrounded by a dewar of liquid helium at four degrees above absolute zero, so the sensing area in the heart of the cylinder is superconducting and sensitive to very small changes in current. The tall assembly above it allows the technician to raise, lower, and turn a sample manually in the sensing region. The small black cylinder to the right on the desktop is a magnetic coil for demagnetizing samples with alternating fields. This entire operation was completely manual in operation, very labor intensive, and very slow. It did not have any magnetic shielding around it or any device for thermal demagnetization of specimens, which are now standard in all paleomagnetics labs. (Photograph by the author)

any small sample of rock. He had grant money to pay his lab technician to work the day shift, but once I arrived at Woods Hole, I stayed up for 48 hours and then began working in the night shift, sleeping in his spare bedroom by day and working 12 hours straight through the night.

Chuck had one of the first-generation cryogenic magnetometers, which is built around a dewar full of liquid helium at four degrees above absolute zero (4°K). This extremely cold helium bathed a series of sensors, which were so cold that they were superconducting and had almost no resistance to electrons. When a sample was brought into the sensing region of the magnetometer, even the weak field generated by a rock would cause a current to flow in the superconducting electronics of the magnetometer and give a measurement of the direction and intensity of the rock sample's magnetic field. In 1979, this machine was state of the art, with only a few in existence. Most of the older labs had spinner magnetometers, which required each specimen to be spun on a special device for more than half an hour just to get one measurement. But the cryogenic magnetometer got a reading in just a few seconds after the sample was lowered into the sensing region, and it could measure samples that were 10,000 times more weakly magnetized than any earlier-generation magnetometer could.

Why would anyone care about the magnetic field of a tiny cube of rock? Magnetometers were the crucial instrument in a newly developed field of geology called *paleomagnetism*, the measurement of the ancient magnetic field recorded in a rock. Many of the early paleomagnetists—whom some people nicknamed "paleomagicians" because their data often provided amazing and magical answers to tough geologic problems—were able to show by using these tiny magnetic samples how continents had drifted across Earth's surface, sometimes traveling huge distances. Others used the field direction to decipher the behavior of Earth's ancient magnetic fields and from that to deduce something about the properties of Earth's outer core, where that field was generated.

All of these studies worked on the same basic principle. When lava cools below 630°C (1,166°F), its magnetic minerals such as magnetite (Fe_3O_4) and hematite (Fe_2O_3) feel the magnetic field at the time and record it in their crystals. Thus, each igneous rock is full of tiny magnets only a few tens of microns in diameter. Each tiny magnet is a record of the direction of Earth's field at the time the lava cooled. After these rocks are eroded and their magnetic grains are weathered out, they become sedimentary grains of iron oxide, which can collect in the sands, silts, and clays deposited on floodplains, in river channels, and on the ocean floor. As the soft soupy sediment begins to pack and harden, the magnetic grains will align all over again to record Earth's field at the time they formed. Thus, we had two potential ways to get a record of Earth's magnetism: cooling from a lava (thermal remanent magnetization, or TRM) and alignment of magnetic grains in soft, wet sediment (detrital remanent magnetization, or DRM).

Paleomagicians who needed to study the ancient direction of Earth's field as recorded in hard rocks, such as lavas, typically used a coring drill (figure 3.2) to obtain their samples. This drill looks a bit like a chain saw (and it is just as noisy), although instead of the chain of saw teeth, it had a rotating cylindrical bit at the end, tipped with diamonds for cutting rocks. In addition to the driller, another crew member stood nearby pumping a pressurized plastic water jug and forcing water through a hose and inside the spinning hollow drill bit. This action sprayed water all over the place, but also kept the bit and the rock cool through the intense friction and heat of drilling and helped flush out the powdered rock from the drill hole. The driller needed to wear a raincoat, waterproof pants, and goggles because of the coating of fine mud that sprayed all over, as well as ear protection to minimize ear damage.

But my type of research was different from traditional paleomagnetic methods. I used a new technique developed in the late 1960s by one of my graduate advisers, Neil Opdyke, and known as *magnetic stratigraphy*. In this method, we would take numerous samples spanning long, thick sequences of rock (and therefore a relatively long chunk of geologic time). We would measure the samples at each level to see what direction their magnetic fields pointed. Some of the samples

FIGURE 3.2 The author drilling paleomagnetic samples near Gualala, California. The portable coring drill works like a chain saw, except that it spins a rotating 1-inch-diameter cylindrical bit with diamonds on the tip for cutting cylindrical cores of rock. The black hose runs to a water bottle with a hand pump, which keeps water flowing through the bit to cool it off and flush out the ground-up rock. So the driller wears a raincoat, goggles, and ear plugs, and he prepares to get sprayed with a film of muddy water as the drilling proceeds. (Photograph by the author)

might be of "normal" polarity (i.e., pointed in the same direction as the modern magnetic field). But others might have reversed polarity, so that their ancient compass directions pointed south, the opposite of what a compass does today.

Back in the 1950s, paleomagnetists Alan Cox and Dick Doell had shown that Earth's field reverses every once in a while, so that if you had held a compass around 800,000 years ago, it would have pointed to the South Pole. This pattern of reversals was random and irregular, with hundreds of polarity changes over the past 70 million years, so there was a distinctive pattern of both short and long reversed and normal polarity events (figure 3.3). Plotted over time, the field reversals recorded in rocks gave a series of irregular black (the convention for normal polarity) and white (reversed polarity) stripes that resemble a bar code. Indeed, if you have a relatively long sequence of rocks alternating between normal and reversed polarity, you can match that pattern to the global magnetic polarity timescale and date rocks that are millions of years old but previously undatable to the nearest 100,000 years or less. This technique was such a huge improvement over previous dating techniques that by the late 1970s many people were jumping into magnetic stratigraphy as a way of improving the dating of fossils as well.

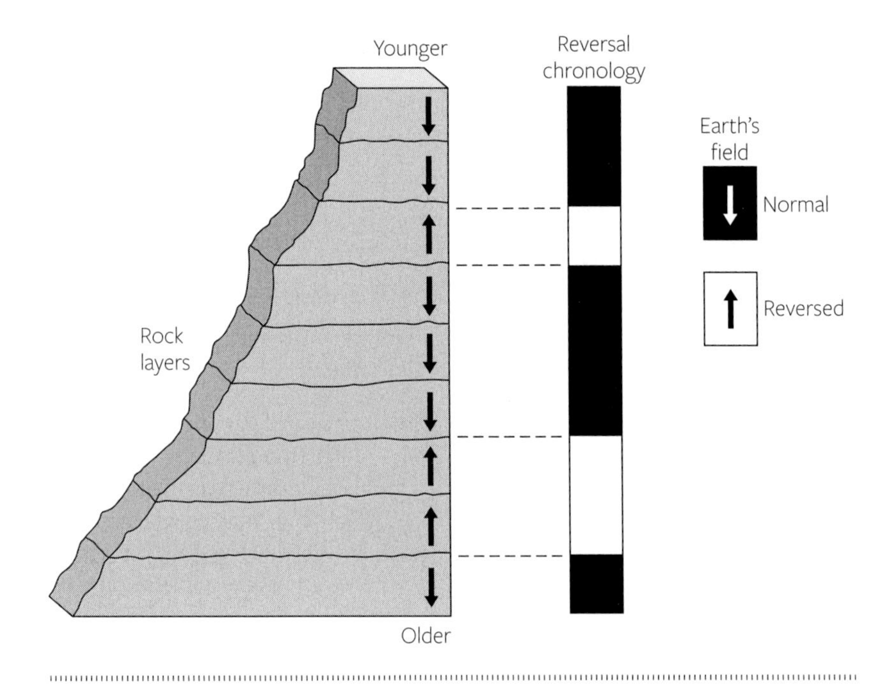

FIGURE 3.3 Standard pattern of magnetic polarity reversals in a thick section of rock.

When I arrived at Columbia University and the American Museum in 1976, magnetic stratigraphy of fossil-mammal-bearing terrestrial beds was just in its infancy. Neil Opdyke and colleagues were about to publish one of the first such studies on the famous late Cenozoic beds of the Anza-Borrego Desert in southeastern California. One of my Columbia predecessors, Bruce MacFadden (now a curator at the University of Florida Museum), had undertaken a dissertation involving just a handful of samples in the late Miocene–Pliocene Chamita Formation near Santa Fe, New Mexico. He had just finished defending that dissertation and had headed for his first job at Yale when I arrived in New York. The student who had arrived at the American Museum one year ahead of me, Steve Barghoorn, had tackled the Miocene rocks of the Santa Fe area just below MacFadden's Chamita Formation for his dissertation. After a summer in the field with my adviser Malcolm McKenna in 1977, our crew of students visited Steve in New Mexico. Steve had been working completely alone out in those hot desolate badlands, but had done a huge amount of sample collecting despite dangerous conditions and working alone. We helped him out a bit, then drove one of the museum trucks he was using back to New York.

When it came time for me to pick a dissertation topic, I could see that paleomagnetism was the hottest new technique around and that magnetic stratigraphy would be an ideal method to help date the rocks and fossils of the Badlands much better than they had ever been dated previously. But I also benefited from new technologies and by learning from my predecessors' mistakes. For sampling in soft sediments, a traditional coring drill was useless because it would pulverize or dissolve the soft sands and silts. Instead, you needed to collect hundreds of small hand samples inscribed with a north arrow on their surface that oriented them with respect to the modern magnetic field (figure 3.4). With so much manual labor, it was foolish to try to collect so many samples alone. Collecting small, oriented, hand samples was a perfect task for a small crew of students, who could thus spread the tedious manual labor among many sets of hands. Barghoorn had taken hand samples with many different orientations and told me that this approach was a mistake—it was very easy to misrecord the orientation of the sample when the surface might be horizontal, vertical, or at any convenient angle.

So I simplified the procedure even further. Because the rocks were soft and easy to scrape with a Stanley pocket plane, chisel, and rock hammer, I cut every sample with a horizontal surface (as determined by a bubble level) on top. This simplification also made it easy for everyone on a crew to collect independently. All they had to do was scrape that horizontal surface, mark a north arrow on the top of their sample with their little compass, break it out in one piece (the trickiest part), and then wrap it and label it. Nobody had to take any measurements

FIGURE 3.4 (*A*) Oriented hand samples work better in soft sediment because the coring drill would pulverize or dissolve the rocks. A horizontal surface has been scraped on top of this block (using a bubble level), and then the modern north direction is marked on that surface before it is removed from the ground. (*B*) Collecting samples requires climbing steep, crumbly slopes and cliffs and working in the hot sun hour after hour as you scramble to keep your footing. This field crew from 1986 was (*top to bottom*) Dana Gilchrist, Allison Kozak, and Kecia Harris. (Photographs by the author)

of the orientation of the surface because it was horizontal every time. Nor did anyone have to record this orientation for each sample individually, which Barghoorn had realized was the biggest source of errors in data collecting. The only tricky part was keeping track of the sample number. I was in charge of figuring out where to measure the rock thickness and run the sampling traverse up the slopes, mark the sampling levels on the rock for the crew to follow, and keep track of where each sample came from (usually including a photo of each site with my field assistant working on it). If the crew was working slowly and I had finished all the section measuring, data recording, and photographing each site, I'd jump in with my own set of tools and sample the rest of the section from the top down so that we'd finish together.

The other scientific breakthrough was the cryogenic magnetometer, which measured a sample in seconds rather than in hours like the old spinner magnetometers. Combining all these efficiencies in sampling with much greater speed and accuracy in measurement, I was able to collect and measure several thousand specimens for my dissertation, whereas MacFadden just a few years earlier was able to measure only a few hundred for his dissertation. Thus, I could tackle a really big project, such as measuring, sampling, and analyzing the paleomagnetism of dozens of sections in the White River Group.

In the 1979 field season, my crew of two students and I finished sampling six major collecting areas in Nebraska and Wyoming, and the results came back excellent. This outcome gave me leverage to ask the geology program at Columbia for a small grant to do another field season. I also applied for a dissertation improvement grant from the NSF, but the reviewers said the project was too ambitious and couldn't be done! So with my shoestring budget, two of my students at Vassar (where I was teaching at the time) and I went out in the summer of 1980 and did what the NSF reviewers said was impossible: I sampled all the major White River sections from western North Dakota through most of the Big Badlands of South Dakota, plus White River outcrops in Wyoming, Nebraska, Montana, and northeastern Colorado. I trimmed the samples, then spent August 1980 doing another long month of 12-hour night shifts getting all the samples analyzed at night and sleeping by day. Working graveyard shift, I became a recluse, seeing little of people or daily activities, but enjoying the walk along the Cape Cod roads from Chuck's house to the lab each twilight and back again at dawn. (I did manage to see the second *Star Wars* movie, *The Empire Strikes Back*, in the theaters before I kicked my sleep cycle over to the night shift.) By the fall, all my results were done, and I focused on writing up my dissertation, finishing the biostratigraphy, and teaching part time at Vassar again.

And so we return to the start of this tale, the summer of 1983. Even though my crew was fantastic and our morale excellent, we had our share of adventures and adverse conditions. While working near Flagstaff Rim, Wyoming (figure 3.5) (Emry's old dissertation area and a classic Skinner locality), we got trapped by a huge overnight thunderstorm that turned the clay-rich soils around us into gumbo. We crammed ourselves like mummies in my little Eureka Timberline tent. It was claustrophobic, but to step outside to the latrine meant scraping muck off our boots for half an hour, so we all got as comfortable as we could and did a *lot* of reading. We simply had to wait until the gumbo clay dried enough that we could hike over to the rim and do our sampling, and it had to be even drier to drive our truck out of that place.

Out in the John Day beds, we had secured permission from the landowner to sample one extremely thick but crucial section, but he would allow us on the land for only one day. Thus, we started before dawn, worked like maniacs without anything but a quick lunch break to unload our heavy packs filled with hundreds of samples, and worked until it was dark, when we finally finished. More than 1,000 samples in just under 11 hours! Boy, our backs, hands, and arms were sore after that grueling marathon!

FIGURE 3.5 My crew camped at Flagstaff Rim, July 1983. We have just unloaded my 1980 Dodge pickup, which the crew nicknamed "Jabba the Truck," and set up the folding table, but the tent and the rest of camp are not yet ready. Rob Lander (*left*) and Annie Walton (*right*) are preparing for a long, hard day of work. Beginning the next morning, we were trapped in our tent on this spot for days when the rains turned the ground into gumbo, and we couldn't even walk around, let alone work or drive out. (Photograph by the author)

But the most scary incident occurred on June 25, 1983, in the Big Badlands. We had just stocked up on supplies at the store, so the truck contained enough food, water, and ice in the cooler to last four or five days. We drove out shortly after dawn, determined to get started and get a lot done before the midday heat made our tasks miserable. First, we stopped at Cottonwood Pass, a legendary Frick collecting area on the south side of Sheep Mountain Table, where we collected 90 samples in a few hours. Then we headed to the west side of Sheep Mountain Table, a long way from the nearest dirt road and as isolated a place as there is in the lower 48 states. We were trying to find the original sections of John Clark's members of the Chadron Formation, which he called the "Ahearn," "Crazy Johnson," and "Peanut Peak" members. (Apparently, there had once been a farmer named Johnson who tried to raise peanuts on one of the sage-covered buttes, which certainly qualifies as a crazy act.) Following a rough dirt track in my big Dodge pickup (which the crew had nicknamed "Jabba the Truck" in honor of the huge, sluglike character from *Return of the Jedi*), we made slow progress (see figure 3.5). Soon the dirt trail vanished, and we were driving cross-country over the grass and rocks to locate this godforsaken place when the engine died suddenly. We tried restarting it, to no avail. Rob (who was a good mechanic) and I looked under the hood and crawled around under the truck, but couldn't find the problem, although we suspected it was a fuel pump. It was getting late in the afternoon, so there was no choice. We decided that Annie and I would walk all the way back to find help, while Rob and Allison set up the tent and stayed put with the truck and all the supplies we had left. Curious cattle kept them company for the entire time we were gone.

Annie and I loaded up with canteens and essential food and clothing, then set out to hike back across our old tracks. That spring had been one of the wettest in recent memory, so the grass was long, there were many ponds, and the mosquitoes were unbelievable! We became two moving clouds of mosquitoes, constantly waving our arms as we walked to keep them from landing on us, and hoping that our hats and shirts and jeans would protect most of our skin surfaces. Not even Off or Cutter could keep these bugs at bay! As we walked, however, the mosquitoes managed to make it through some of our defenses. When I took my shirt off that night, solid mosquito bites ran along the outlines of my canteen and backpack straps!

We walked on and on for hours. At one point, Annie saw me step right over a rattlesnake that I could not see because I was waving so hard to keep the mosquitoes off me. The snake must have been equally startled because it didn't have time to coil up or buzz its rattle or attempt to bite through my boots or jeans. Finally, we reached the nearest dirt road, which runs along the east side of Sheep Mountain Table, and tried to hitch a ride. Nobody came for a while, until this beat-up old pickup drove by with two young Lakota men inside, headed south toward the Pine Ridge Reservation. This was the opposite direction we needed to travel to get to Interior, South Dakota, and the Badlands headquarters, but we were in no condition to be picky, so we rode in the cluttered bed of the pickup until we reached Sharps Corner, where they dropped us off. Then we had to hitch a ride with another car full of Lakotas headed east through Kyle and Potato Creek to the junction of Highway 44, and finally a third hitch north with yet another car driven by Lakotas got us into Interior late that Saturday night. Of course, by that time it was too late to find anyone to help, so we rented a little trailer, then visited the only service station in Interior the next morning.

Because the owner and mechanic didn't want to drive out to find our stranded truck until he had a new fuel pump for a 1980 Dodge pickup, which he didn't have in stock, we spent the entire Sunday reading, hanging out at the park service restaurant and the visitor's center, and worrying about how Rob and Allison were doing. (This was long before the age of cell phones, not that there's much cell reception out there even now.) On Monday morning, a fuel pump came in from Rapid City, and the mechanic took us to our breakdown site in his own big four-wheel-drive pickup because of the rough roads. We almost couldn't find our own tracks after two days because the grass had grown back and our tire tracks were almost invisible. Finally, we found Rob and Allison, and, boy, were they happy to see us! Rob and the gas station owner got to work, but once the fuel pump had been replaced, Jabba still didn't want to start.

There was no choice—the mechanic was going to have to jerk us out of the Badlands with his pickup. I took the wheel of my own wounded truck, but I didn't have much control because without engine power, the brake and steering hydraulics were next to useless. We went up one ridge after another and down ravine after ravine, while I held on to the wheel for dear life and pounded the nearly useless brake pedal with all my might to try to control this huge, disabled hunk of metal. Finally, we reached the regular highway that would take us back to Interior, but then I faced a different problem: without brakes, I couldn't slow down easily and prevent myself from gaining on the vehicle that was towing me. Every time I'd creep up and cause the stretchy nylon rope holding us together to touch pavement at high speed, it would melt! Then we'd have to stop, knot the two halves of the rope back together again, and try to keep going.

Finally, we straggled into Interior late on Monday night, and Rob and the gas station owner postponed looking at the truck until the next morning, when the light was better. It turned out that the fuel pump was not the problem after all, but instead the fuel line for an auxiliary fuel tank I had installed to give the truck more range in backcountry. Its fuel line was connected to the main fuel line with a little valve that was controlled by a wire from the dash switch. During our bumping along the backcountry on Saturday, we had somehow jerked this wire loose from the dash switch, so even though there was plenty of gas in both tanks, the valve was

FIGURE 3.6 Sampling in the Ahearn Member of the Chadron Formation west of Sheep Mountain Table in the Big Badlands in the summer of 1983. The crew and I had made a second and successful attempt to reach the region after our first attempt failed. Annie Walton is visible in the foreground and Rob Lander in the distant background, taking samples. (Photograph by the author)

stuck in the middle and no gas from either tank could reach the engine. Once the mechanic had fixed the problem (and charged us much less than I expected, given how much time, labor, and gas we had consumed), we were forced to head west to the John Day beds because we'd lost four days and had a schedule to keep.

At the end of the summer field season in mid-July, we made yet another attempt to reach Peanut Peak and the type Chadron exposures west of Sheep Mountain Table. As noted in chapter 2, we luckily stumbled upon John Clark at a campfire talk, and he told us that the roads out of Scenic around the north side of Sheep Mountain were much better. Sure enough, we followed those roads and got there with no problem on our second try. We collected our samples (figure 3.6), and later that summer I ran them: they all were scientifically worthless. They were heavily stained and impregnated by brick-red iron hydroxides, making them completely overprinted with modern magnetic fields and useless for paleomagnetism. Thus, the curse of Crazy Johnson had struck yet again, and all that work and danger accomplished nothing. However, such failures are part of science, and I'm happy to report that most places we sample are much less dangerous to reach and usually yield good results to the magnetometer.

Magnetic Flip-Flops

So why do we spend so much time, work so hard, and risk our lives to collect small, oriented pieces of rock? The answer, as mentioned earlier, concerns dating. Prior to the late 1970s, there was no way to date the rocks of the White River Group precisely or to determine the fossils' numerical age. We knew that there were different assemblages of fossil mammals in the Chadronian, Orellan, and Whitneyan land mammal ages (see figure 1.3), but the numerical age of those rocks in millions of years was a mystery. The Chadronian had traditionally been considered early Oligocene, the Orellan middle Oligocene, and the Whitneyan late Oligocene. Unfortunately, this correlation was based on relatively few mammals in common between North America and the marine sections in Europe that had defined the basis for early, middle, and late Oligocene. Anything we could do to improve the dating of these rocks and fossils had tremendous potential to refine our interpretations about how long these creatures had lasted and how the changes in these mammals through time might be interpreted. Most of all, if we could get a well-calibrated magnetic pattern, we could match the White River magnetozones to the standard marine stratigraphy of the world's oceans and find out exactly what land animals, plants, and climate had done when the major climatic changes had happened elsewhere in the world (especially in the oceans).

This endeavor indicated the huge potential of magnetic stratigraphy in the White River sequence. So many excellent, well-exposed sections, but the only correlation between them was based on the general similarities of the mammal fauna and on a few widespread marker layers. With the detailed Oppelian range-zone biostratigraphy I was compiling, I could improve the general correlation between each section, from Montana, North Dakota, and South Dakota through Wyoming and Nebraska and down to Colorado and Texas. With the addition of the magnetic zones in each section, I could put little "bar codes" on each section and match up the "stripes" along the entire outcrop of the White River Group, which would give a precision of correlation to the nearest meter, and each magnetic reversal event would be datable to the nearest 100,000 years.

And that's what I did in the summers of 1979, 1980, 1983, 1987, and 1988. By the time I finished, I had magnetic "bar codes" on virtually every important White River section that had ever been collected, and a general pattern emerged (figure 3.7). There was a normal polarity zone at the base of the Orellan in nearly every section I sampled, although the sections near Lusk and Douglas, Wyoming, had a thin Orellan reversed polarity zone right below the normal that was not represented farther to the east. The entire rest of the Orellan and the early Whitneyan was reversal in polarity. Then there was a short normal in the middle Whitneyan, but a reversal again in the uppermost Whitneyan. Finally,

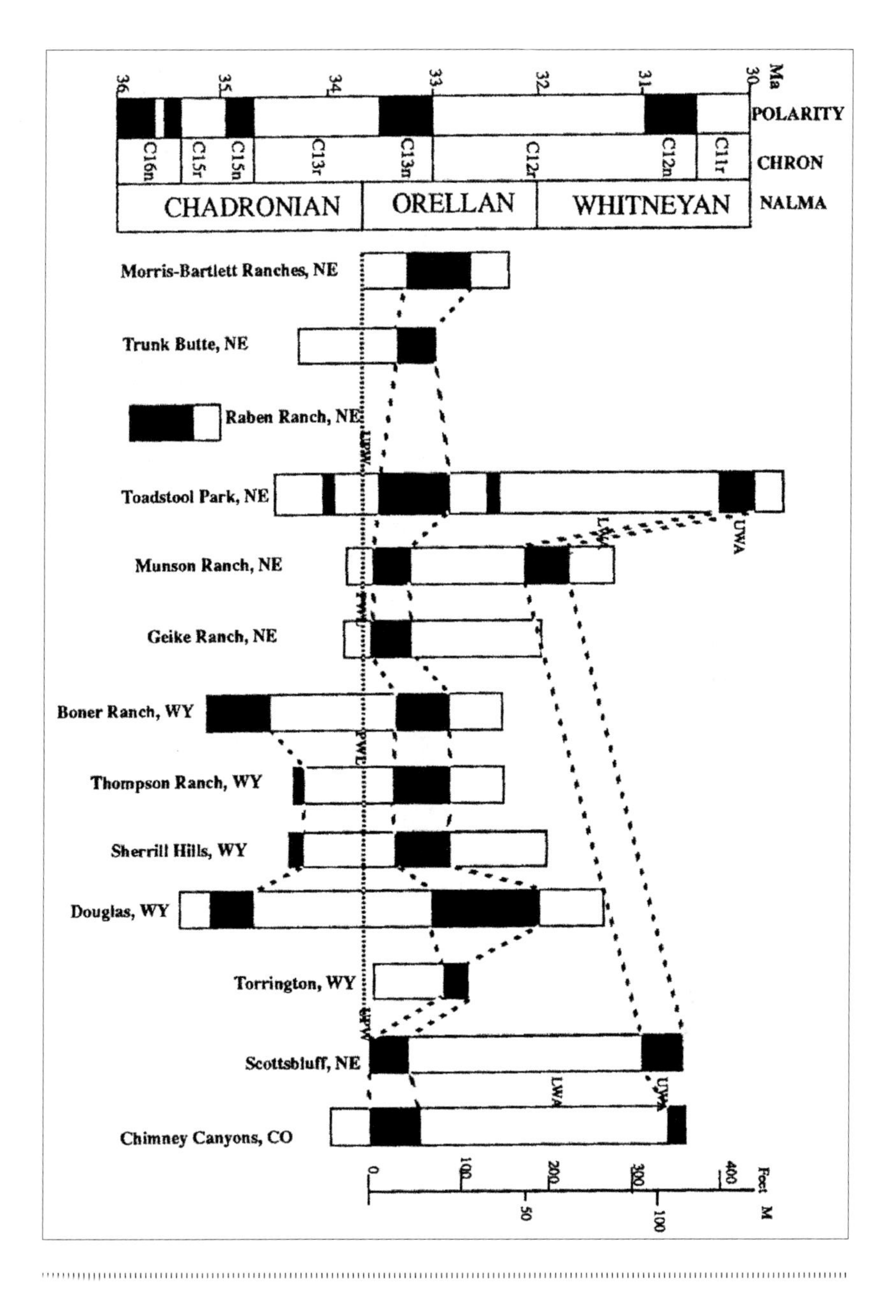

FIGURE 3.7 Correlation of the magnetic pattern in the major sections of the White River Group. The black bars indicate normal polarity, and the white bars show reversed polarity. The magnetozones are aligned along marker ashes (PWL = Persistent White Layer; UPW = Upper Purple White layer; LWA = Lower Whitney Ash; UWA = Upper Whitney Ash) or along the biostratigraphic boundary between the Chadronian and the Orellan. (NALMA = North American land mammal "ages.") Even though each section has different thicknesses and sedimentation rates, the relative thicknesses of the polarity zones is consistent and matchable from one section to the next.

the base of the overlying Arikareean was normal in polarity, and that's as far as I had sampled by 1988.

Thus, I could correlate each section by two independent means: biostratigraphy and magnetic stratigraphy. The two systems matched up quite nicely, so I used this web of correlation to answer another question that became the central hypothesis of my dissertation. Back in the early nineteenth century, paleontologists had noticed the similarity of faunas in many regions and made the assumption that similar fossils denoted similar ages. But how accurate was this assumption? The great biologist and geologist Thomas Henry Huxley—known as "Darwin's bulldog" for his defense of evolution—had posed this very question in his 1862 Presidential Address to the Geological Society of London. He argued that just because we see a similarity in order of fossil faunas (homotaxis), we cannot assume that each similar fauna is the same age.

From those days until the 1970s, there were not many ways to test the synchroneity of biostratigraphic events. When paleomagnetism came along in the 1970s, however, it provided a perfect test. A magnetic reversal takes only 4,000 to 5,000 years to complete, which is a geological instant for rocks that are 30 to 40 million years old. In the outcrop, a reversal is usually a sharp boundary between one sampling level and the next, so it provides a perfect time plane. I compared the biostratigraphic events I had identified in the White River mammals with my magnetic time planes, and in no case did a biostratigraphic horizon span a polarity boundary. Thus, within the resolution of my study, I had shown that the basic assumption of homotaxy in mammalian biostratigraphy worked. I arranged my entire dissertation project to be a test of this fundamental question. In the summer of 1982, I argued this very point in one of my first professional talks based on my dissertation at the North American Paleontological Convention in Montreal.

Radioactive Revolution

Although my magnetic stratigraphy and biostratigraphy were highly consistent with each other and provided a nice correlation among my sections of rock, there was still a more important question to answer: How did my correlated sections of rock match up with the global timescale? This question could be answered in only two ways, and neither had been successful so far. In some parts of the world, terrestrial beds with diagnostic land mammals were interbedded with marine strata that contained diagnostic marine fossils. In such places, concepts such as Eocene and Oligocene (which were defined on marine rocks and fossils in Europe, especially Italy, France, and Belgium) could be extended to tell us what beds in North America correlated with the Eocene or Oligocene. Unfortunately, such interfingering between marine and nonmarine rocks simply didn't occur in Chadronian, Orellan, and Whitneyan in North America, so it was anyone's guess where the North American land mammal ages fit in the European sequence.

The only alternative was another technique that had been available only since the late 1950s: *radiometric dating*. This system of numerical dating (formerly but incorrectly called *absolute dating*) allows us to talk about the ages of geologic events in numbers, such as thousands, millions, and billions of years before the present, rather than in the relative terms such as "late Eocene" or "middle Oligocene." The dating technique is based on the principle of radioactive decay of certain unstable elements, such as potassium-40, rubidium-87, and uranium-235 and 238. In the case of relatively young rocks such as those of the White River beds, the only useful method was potassium-argon dating, where radioactive potassium-40 changes to argon-40 with a known rate of decay. If you can take a crystal of a mineral in a volcanic ash that is fresh and unweathered and measure the ratio of the radioactive parent potassium-40 to its decay product, argon-40, that ratio will give you a numerical age.

When potassium-argon dating first became practical in the late 1950s, it was in huge demand by nearly every geologist on the planet because it was the only practical means of dating most rocks that were younger than 600 million years old. (By contrast, rubidium-strontium dating and both kinds of uranium-lead dating decay too slowly to be used except on very ancient rocks at least 1 billion years old.) I remember hearing stories about the battles in scientific circles over who would control this valuable—and at the time expensive and rare—technique that gave almost any geologist a numerical age better than any that had been obtained earlier. At the USGS in Menlo Park, California, Brent Dalrymple (a 1959 graduate of the institution into which I finally settled, Occidental College, or "Oxy") had become one of the first experts on potassium-argon dating, and his skills and laboratory were much in demand. Fortunately, the top officials at the USGS paired him up with Allan Cox, and their combined skills were crucial in developing the early magnetic polarity timescale and eventually in confirming plate tectonics.

At the same time, Jack Evernden and Garniss Curtis developed the potassium-argon geochronology lab at Berkeley. It was state of the art at the time, and everyone wanted to have their dates run there. Although the Berkeley lab had a huge number of potential geologic problems to work on, it, in collaboration with Berkeley paleontologist Don Savage, made a point of analyzing samples of volcanic ash from as many different North American land mammal ages as possible. Morris Skinner had collected many of these ash samples and had revisited his classic localities just to get more ash dates for whatever lab would run them. In

1964, Evernden, Savage, Curtiss, and Berkeley grad student Gideon James published a classic paper that gave the first decent potassium-argon dates on land mammal faunas.

The striking thing about these early dates was that they all made sense. They all were consistent with the sequence of mammalian faunas that Matthew and Osborn had worked out back in the 1890s, with no exceptions. This consistency by itself showed the robustness of the North American land mammal age sequence and was confirmation that mammalian biostratigraphy worked well. Nevertheless, there were many areas where Evernden and colleagues didn't have good age control or where there were gaps in the sequence with no dates available (such as for the entire Orellan and Whitneyan) or where the dates were pretty rough with big plus or minus error estimates. But that was the best that could be done with the technology and samples available in the early 1960s.

Every mammalian paleontologist after 1964 calibrated their mammalian timescales and placed rough numerical ages on key events based on the dates published by Evernden and his colleagues that year. It wasn't very precise, but it was better than anything that came before. During the 1960s, the 1970s, and even the 1980s, more and more potassium-argon dates were added to the matrix, so the correlations and age estimates became better and better. The classic Cenozoic mammalian timescale volume, edited by my former undergrad adviser Mike Woodburne and published in 1987 after 15 years of delays, was based entirely on these old dates, and everyone had no choice but to use these correlations. When I published my first papers on my White River magnetic stratigraphy, I correlated the magnetic pattern to the global timescale based on these 20-year-old dates and matched the long reversal in the late Orellan–early Whitneyan to the long magnetic Chron C10r of the magnetic timescale (figure 3.8). There was nothing else available back then except the few dates that Evernden and the others had run based on Morris Skinner's samples from Flagstaff Rim.

It's About Time

Once I had my potassium-argon dates and magnetic stratigraphy and biostratigraphy, I found myself caught up in a larger dispute: the calibration of the global timescale. We see the standard geologic timescales all over the place (see, for example, figure 1.3), but we seldom think about how they are constructed or what they are based on. In reality, each timescale is a complex interaction between radiometric dates from volcanic layers that calibrate some local biostratigraphy plus many other components, such as magnetic stratigraphy. When the

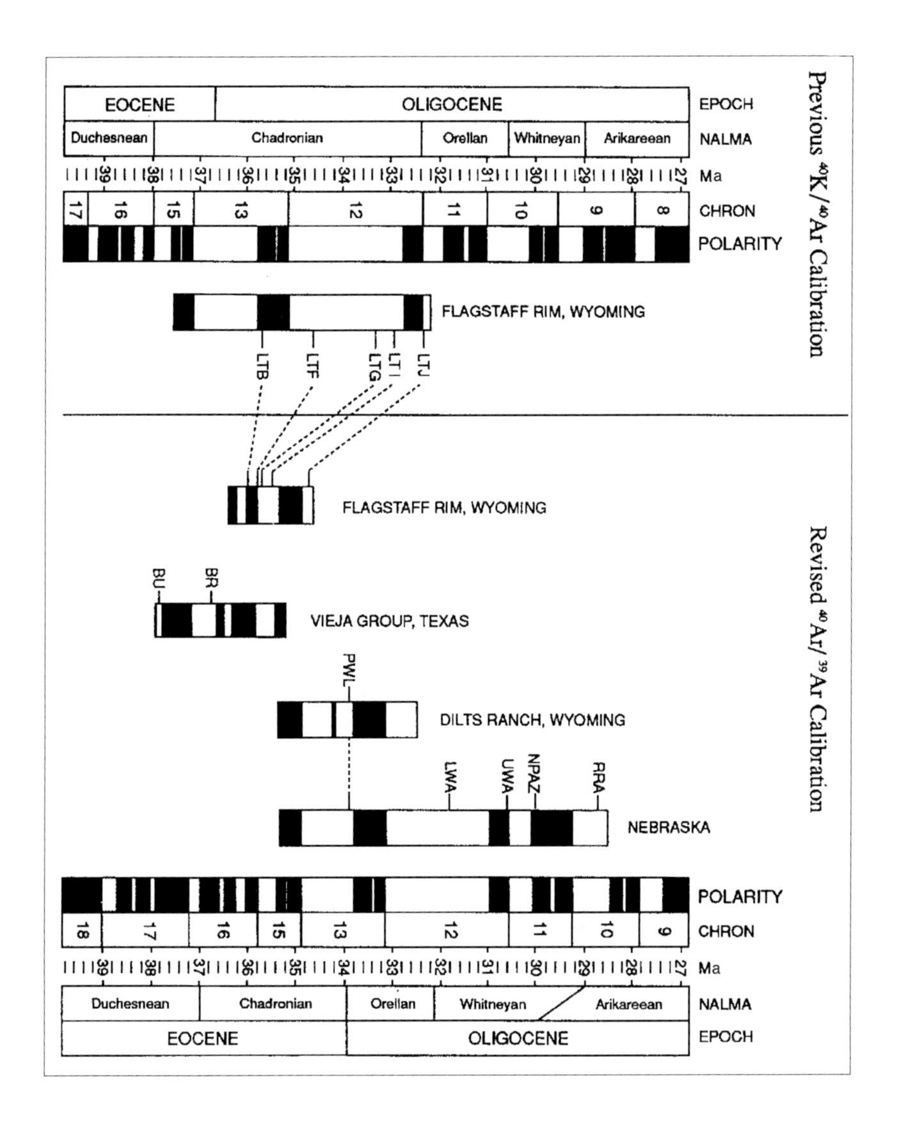

FIGURE 3.8 Correlation of the magnetic timescale with the White River sections, using the older potassium-argon (^{40}K/^{40}Ar) dates of Jack Evernden and colleagues (1964) on the top and Carl Swisher's newer argon-argon (^{40}Ar/^{39}Ar) dates on the bottom. Some correlations are radically rearranged, especially with the change in the Flagstaff Rim dates. PWL = Persistent White Layer; LTB, LTG, LTI, LTJ = Flagstaff Rim Lone Tree Gulch ashes B, G, I, J; BR = Bracks Rhyolite; BU = Buckshot Ignimbrite; LWA = Lower Whitney Ash; UWA = Upper Whitney Ash; RRA = Roundhouse Rock Ash; NPAZ = Nonpareil Ash Zone; NALMA = North American land mammal "ages." (From Prothero and Swisher 1992)

biostratigraphic concepts of the Eocene and Oligocene were first set up starting in 1833 with Charles Lyell, who named the Eocene, they were based on the marine faunas of classic localities in Italy, France, Belgium, and Germany. Unfortunately, these classic Eocene marine fossils (mostly mollusks) were often shallow-water species that were hard to correlate around Europe and virtually useless in correlating the Eocene or Oligocene elsewhere in the globe.

By the late 1960s, however, another new source of data had emerged: the biostratigraphy of marine microfossils, especially amoebalike fossils such as the foraminifera. These tiny plankton occur by the millions in a few liters of seawater and are so abundant and rapidly evolving that they have become the standard index fossil for most of the Mesozoic and all of the Cenozoic. Starting in the late 1960s, deep-sea cores full of these microfossils were drilled from many parts of the world's oceans, and this marine planktonic timescale soon became the global standard. In addition, many of these marine cores were perfectly undisturbed, so scientists such as Neil Opdyke were able to sample them for magnetic stratigraphy. Thus, the deep-sea cores with their planktonic microfossils and magnetics were the ultimate standard for telling geologic time in the oceans.

How did these cores match the classic concepts of Eocene and Oligocene in Europe? Unfortunately, the classic "type" sections in Europe described by the first stratigraphers in the early nineteenth century consisted of very shallow marine strata that usually didn't contain any well-preserved diagnostic microfossils. They occasionally did yield a few, and on this basis micropaleontologists such as Bill Berggren at Woods Hole Oceanographic Institute were able to calibrate how the Eocene and Oligocene and other Cenozoic stages and ages in Europe matched with the global planktonic microfossil/magnetic timescale. In classic papers in 1971 and 1972, Berggren laid down all the evidence for these correlations and gave the timescale a firm foundation.

Assigning numerical dates to the global magnetobiochronology was a different matter, however. Most of these deep marine rocks contained no volcanic ashes or lava flows, so potassium-argon dating was impossible. Even the shallow marine rocks seldom produced good dates. The only suitable mineral for dating seemed to be the green clay known as glauconite, which grows on the sea bottom, but does contain enough potassium to be dated. Some stratigraphers, including Gilles Odin in France and Dennis Curry in England, had generated a new timescale based on glauconite dates. According to them, the Eocene–Oligocene boundary was only 32 million years old, far younger than anyone had thought.

The other way to approach the problem was to use high-temperature volcanic ash dates from land sequences that calibrated certain points on the magnetic timescale. In this regard, the potassium-argon dates published by Evernden and others in 1964 were crucial to calibrating my White River magnetic stratigraphy. These dates were part of the growing nexus of other high-temperature dates that seemed to support an age of about 36 to 38 million years for the Eocene–Oligocene boundary. During the 1980s, there was a major battle over the timescale, with Berggren and others on the one side arguing for the superiority of high-temperature volcanic ash dates, and Odin and Curry on the other defending their glauconite dates. Berggren and others were able to show that glauconite dates are highly suspect because they readily leak the daughter product (argon) and form on the seafloor, sometimes millions of years after the beds were originally deposited.

Other groups of scientists joined the fray. Some, such as Bill Lowrie at ETH in Zurich, and Walter Alvarez at Berkeley, took some new dates and completely stretched the timescale to fit them, giving a totally different age estimate for the Eocene–Oligocene boundary. Berggren and Dennis Kent (also a Columbia faculty member and a member of my dissertation committee) showed that this practice was illegitimate because the magnetic timescale was based on the magnetic "stripe" patterns of the deep seafloor spreading ridges and could not be arbitrarily squeezed down or stretched out like a huge rubber band. Those seafloor magnetic "stripes" were formed when the ocean floor was slowly generating by spreading in the midocean ridges, a discovery that clinched the hypothesis of plate tectonics in 1963. It was extremely unlikely that the seafloor had rapidly sped up or slowed down its spreading rate, so the wild fluctuations advocated by people such as Lowrie and Alvarez or Gordon Ness, Shaul Levi, and Richard Couch (1980) were unlikely.

And so by late 1980s there was a general consensus based on the timescale published by Berggren, Kent, and Flynn in 1985. Most geologists were unconvinced by Odin and Curry's unusually young glauconite dates and agreed that the Eocene–Oligocene boundary was about 36.5 million years in age.

Topsy Turvy

I remember it as if it were yesterday. In the winter of 1988–89, I was organizing a big Penrose Conference to be held in Rapid City, South Dakota, in the upcoming summer. A Penrose Conference is an experience unlike anything else in the profession. In contrast to most other geology meetings, it is an invitation-only workshop focused on a relatively narrow topic, intended to bring together all the best specialists on the topic to workshop, argue, and try to hammer out an understanding without the strict limits of a schedule of talks or competing sessions. In this case, I was trying to bring together all the scientists who were familiar with the marine and terrestrial Eocene and Oligocene rocks of the world and get them to talk to each other. Most had never even read each other's work, let alone met each other because the marine micropaleontologists tended to go to their own meetings and the vertebrate paleontologists to their own. Only a Penrose Conference with its broad reach, loosely structured format, and focused topical approach could bring these disparate geoscientists together and see if they had common ground.

This particular Penrose Conference had an all-star cast. There were 60 in attendance, including many giants of marine geology and micropaleontology, such as Bill Berggren, Marie-Pierre Aubry, Jim Kennett, Brian McGowran, Gerta Keller, Jim Zachos, Joe Hazel, and Ken Miller. These micropaleontologists knew microfossils like the back of their hands and were instrumental in calibrating the global climatic record. Many of them were at oceanographic institutes and were familiar with ocean research vessels, but had never spent much time in the continental interior. None had been to South Dakota, so we ran a special side trip so they could see Mount Rushmore. At the middle day of the five-day meeting, we organized a field trip for the entire group through the Badlands and over to the Hot Springs Mammoth Site, so we could get out of dark conference rooms and let people clear their heads and collect fossils. In fact, most of the micropaleontologists had never collected a vertebrate fossil in their lives or walked in real badlands, so this experience was eye-opening for them.

This Penrose also included the major names in vertebrate paleontology, including my former adviser Malcolm McKenna as well as Bob Emry, Mary Dawson, John Flynn, Dick Tedford, John Storer, John A. Wilson, Howard Hutchison, Phil Bjork, Richard Stucky, and Spencer Lucas. We also had a good representation of foreign mammalian paleontologists, including Jeremy Hooker from the British Museum, Jean-Louis Hartenberger and Serge Legendre from Montpellier in France, Kurt Heissig from Germany, and Ewan Fordyce from New Zealand, who worked on early whales. There were major players in paleobotany, including Greg Retallack from Oregon, Jack Wolfe from the USGS, and Margaret Collinson from England. Paleontologists who worked on marine mollusks were also there—John Armentrout, Buck Ward, David Dockery, and Luc Dolin. We also had many stratigraphers and sedimentologists who worked on the Eocene and Oligocene—Emmett Evanoff, Dan Garcia, Debbie Hanneman, Bruce Hanson, Jan Hardenbol, Gary Peterson, Jim Swinehart, Peter Thompson, and Dennis Terry. Two of the specialists in radiometric dating, Bill McIntosh and Carl Swisher, gave us geochronological expertise. Lisa Cirbus Sloan modeled Eocene climates on a Cray supercomputer. All in all, we had the tremendous diversity of scientific expertise needed to tackle a tough problem requiring interdisciplinary approaches.

The Penrose format is designed to maximize the time for interaction and discussion. Everyone who is accepted must give a presentation and contribute, so there are no passive observers. Each presentation is loosely scheduled, so that there should be plenty of time for good questions and discussion before moving on to the next speaker. Each day started off with a keynote speaker, who was supposed to focus our attention on key problems. The entire Penrose was an all-expenses-included, prepaid package deal, including room and all meals together at the same hotel (in this case, the Holiday Inn at Rapid City), the use of the conference room, and the field trip at midmeeting. All in all, it is an extraordinary scientific experience. Penrose Conferences often yield major scientific breakthroughs that can never happen at typical scientific meetings with their hurried pace and huge number of tightly scheduled talks that conflict with one another in multiple sessions.

A few months before the meeting, I received The Phone Call. My good friend Carl Swisher was calling with some startling news. By the time I had hung up, I knew that most of the stuff I'd published on the timescale was out of date, and I'd have to redo much of my work since the early 1980s.

I had first met Carl at that time when he was doing his master's thesis on the Arikareean of Nebraska. He had hosted me at his house in Lincoln while I was studying the University of Nebraska collections. Carl had since then gone on to Berkeley, where his research had moved from mammals to the newest form of radiometric dating, the argon-40/argon-39 dating technique. This method was an offshoot of traditional potassium-argon dating, but was a big improvement on that older method. Instead of measuring a solid parent material (potassium) and a volatile gas daughter product (argon), this method used another isotope of argon to estimate the original potassium concentration. Thus, there was no problem with the daughter's relative leakage because both parent atom and daughter product were the same element.

Argon-argon dating had other huge advantages. In one technique, the geochronologist takes a crystal and slowly heats it in steps, measuring the argon released at each step. If the crystal has leaked or been contaminated, the first measurements (from the outer layer of the crystal, which is most likely to suffer weathering) might give a spurious ratio. As the heating continues, the argon is released from the uncontaminated, tightly closed lattice in the center of the crystal, which gives not only a much more reliable age, but also a method of recognizing contamination and eliminating it.

Carl and the Berkeley lab had a different, equally cool technique. They handpicked only the freshest individual crystals of minerals from a volcanic ash and then mounted each little crystal on a rotating stage. Then the lines to the mass spectrometer were pumped down, creating a vacuum. Each individual crystal was zapped by a laser, releasing all its argon to the mass spectrometer, which sorted and measured the released argon gas (figure 3.9). By doing dozens of individual crystals from a single ash, scientists such as Carl could figure out which dates clustered to give an accurate age and which individual crystals disagreed and were probably contaminants.

FIGURE 3.9 The argon-argon dating technique: (*A*) in the vacuum chamber, a laser zaps each individual crystal and releases its argon as an ionized gas into the vacuum line of the mass spectrometer; (*B*) the ionized gas then travels through a vacuum tube influenced by a powerful eletromagnet, where charged ions of different mass are deflected as they pass by an electromagnet. Those with greater mass are deflected less, so when they reach the sensing region at the other end, they are sorted by mass, and isotopic ratios can be calculated. (Photographs courtesy P. Renne and A. Deino, Berkeley Geochronology Center)

Carl had worked on White River and Arikaree rocks back in his Nebraska days, so he was already interested in the problem of their age and had collected his own samples of the previously undatable ashes from Nebraska and Wyoming. What he found was shocking. Many of the ashes from the Chadronian beds at Flagstaff Rim that Evernden and his colleagues (1964) had used were badly misdated. For example, Evernden and the others had dated the entire Flagstaff Rim section (see the photograph that opens this chapter) between 32 and 37.5 million years in age. Carl's newer, more precise ages had eliminated the problem with contamination of the old potassium-argon dates, and he found that the same section to be only 34.5 to 37.5 million years ago. The biggest change was the date of Ash J, which went from 32.4 million years by the old potassium-argon dates to 34.7 with argon-argon dating (see figure 3.8).

These new numbers, plus the first-ever dates on the Orellan and Whitneyan, which had previously been undatable by the old techniques, completely threw me for a loop. The long interval in the late Orellan and Whitneyan that I'd once correlated with magnetic Chron C10r was actually the much longer Chron C12r (see figure 3.8), two whole magnetic chrons older than I'd been saying for the entire 1980s. When I rearranged everything to fit the new correlations, the Orellan had expanded from 1 million to 2 million years in length, and the Whitneyan went from 1.5 million to almost 3 million years long.

The biggest shock is what happened to the Chadronian. The old potassium-argon numbers (especially Flagstaff Rim Ash J, the most problematic date of all) made the Chadronian 5 to 6 million years long. The new argon-argon numbers reduced it down to 3 million years long (34 to 37 million years ago), so it had shrunk to about half its original length. Even more startling was how the Chadronian correlated with the new global timescale. If my Orellan–Whitneyan magnetics encompassed Chron C12r, then they spanned the early Oligocene (from 34 to 30 million years ago). Thus, the Chadronian had to correlate with Chron C16n to Chon C13r, which made it late Eocene! The long dispute over the Eocene–Oligocene boundary was *finally* solved. It was around 34.0 million years old, not as young as Odin's 32 million years, but younger than the date estimated by Berggren, Kent, and Flynn (1985), 36.5 million years.

This correction may not seem startling to most people, but to a mammalian paleontologist it was as shocking as hearing that Kennedy had been assassinated. Since the late nineteenth century, people had thought that the Chadron Formation and its fossils correlated with the early Oligocene. It had been established that way in the Wood Committee report of 1941, and every paleontologist since then had memorized "Chadronian = early Oligocene" since their grad school days. Scientists tend to be a conservative bunch, unwilling to part with concepts they learned as students, so it takes a great deal of evidence to overcome such inertia.

Instead, the dispute had long been about whether the next oldest land mammal ages, the Uintan and Duchesnean, were late Eocene or also early Oligocene. With Carl's dates, my magnetics, and the global timescale, however, we'd thrown conventional wisdom out the window and undone a century of miscorrelation. The Uintan and Duchesnean were not late Eocene, but middle Eocene; the Chadronian was late Eocene, not early Oligocene; and the Orellan and Whitneyan were early Oligocene, not middle and late Oligocene.

All those textbooks, all those diagrams and charts, all those figures showing the last brontotheres in the early Oligocene—now obsolete because of a few crystals from a volcanic ash and a new technology. Everyone in the older generation of paleontologists had to redo their work and rewire their brains for the new thinking. Many were skeptical at first. I remember getting up in front of the SVP meeting in Austin, Texas, in November 1989 to present this new data, and the harshest critic in the audience was my own former Ph.D. adviser, Malcolm McKenna. He was always on top of such things, but he had previously been burned by bad dates, so he was skeptical until the evidence convinced even him.

As the participants were preparing talks for the Penrose Conference in 1989, I had to redo all my slides just to incorporate the new timescale. We got to Rapid City, and we led the meeting off with our shocking news. It was as if we had formally announced that everyone had to throw out their prepared talk and redo their slides because they were out of date the moment the scientists had arrived. Several scientists, such as John Armentrout and Jack Wolfe, found their talks had become instantly irrelevant, and had to focus on other topics. They had come prepared to talk about their own data that argued for a different age of the Eocene–Oligocene boundary around 34 million years ago, but the conference had opened with new dates that put everyone in agreement.

The new timescale was a huge breakthrough. We no longer had to waste time (as we did through much of the 1980s) arguing about correlations and timescales. We could find a common timescale easily now, and the rest of the conference focused on looking at the details of biotic change in marine animals and land mammals throughout this fascinating transition from greenhouse to icehouse world.

And that was the final benefit of our discovery. Not only had we ended the tiresome battle over different versions of the timescale, but we also had now provided a firm basis for correlating what had happened in climate, extinctions, and evolution in the marine realm with what had happened in the terrestrial realm. No longer was our work merely guesswork. We could now match events up to the nearest magnetic polarity reversal and correlate them with resolution of less than 100,000 years (Swisher and Prothero 1990; Prothero and Swisher 1992). From this point on, the understanding of this interval would never be the same.

Further Reading

Berggren, W. A. 1971. Tertiary boundaries. In B. F. Funnell and W. F. Riedel, eds., *The Micropaleontology of Oceans*, 693–808. Cambridge: Cambridge University Press.

Berggren, W. A., D. V. Kent, and J. J. Flynn. 1985. Paleogene geochronology and chronostratigraphy. *Geological Society of London Memoir* 10:141–195.

Berggren, W. A., D. V. Kent, C. C. Swisher III, and M.-P. Aubry. 1995. A revised Cenozoic geochronology and chronostratigraphy. *SEPM Special Publication* 54:129–212.

Evernden, J. L., D. E. Savage, G. H. Curtis, and G. T. James. 1964. Potassium argon dates and the Cenozoic mammalian chronology of North America. *American Journal of Science* 262:145–198.

Prothero, D. R. 1995. Geochronology and magnetostratigraphy of Paleogene North American land mammal "ages": An update. *SEPM Special Publication* 54:305–315.

Prothero, D. R. 1996. Magnetostratigraphy of the White River Group in the High Plains. In D. R. Prothero and R. J. Emry, eds., *The Terrestrial Eocene–Oligocene Transition in North America*, 247–262. Cambridge: Cambridge University Press.

Prothero, D. R., C. R. Denham, and H. G. Farmer. 1982. Oligocene calibration of the magnetic polarity timescale. *Geology* 10:650–653.

Prothero, D. R., C. R. Denham, and H. G. Farmer. 1983. Magnetostratigraphy of the White River Group and its implications for Oligocene geochronology. *Palaeogeography, Palaeoclimatology, Palaeoecology* 42:151–166.

Prothero, D. R., and C. C. Swisher III. 1992. Magnetostratigraphy and geochronology of the terrestrial Eocene–Oligocene transition in North America. In D. R. Prothero and W. A. Berggren, eds., *Eocene–Oligocene Climatic and Biotic Evolution*, 46–74. Princeton, N.J.: Princeton University Press.

Prothero, D. R., and K. E. Whittlesey. 1998. Magnetostratigraphy and biostratigraphy of the Orellan and Whitneyan land mammal "ages" in the White River Group. In D. O. Terry, H. E. LaGarry, and R. M. Hunt Jr., eds., 1998. *Depositional Environments, Lithostratigraphy, and Biostratigraphy of the White River and Arikaree Groups (Late Eocene to Early Miocene, North America)*, 39–61. Geological Society of America Special Paper no. 325. Boulder, Colo.: Geological Society of America.

Swisher, C. C., III, and D. R. Prothero. 1990. Single-crystal $^{40}Ar/^{39}Ar$ dating of the Eocene–Oligocene transition in North America. *Science* 249:760–762.

Tedford, R. H., J. Swinehart, D. R. Prothero, C. C. Swisher III, S. A. King, and T. E. Tierney. 1996. The Whitneyan–Arikareean transition in the High Plains. In D. R. Prothero and R. J. Emry, eds., *The Terrestrial Eocene–Oligocene Transition in North America*, 295–317. Cambridge: Cambridge University Press.

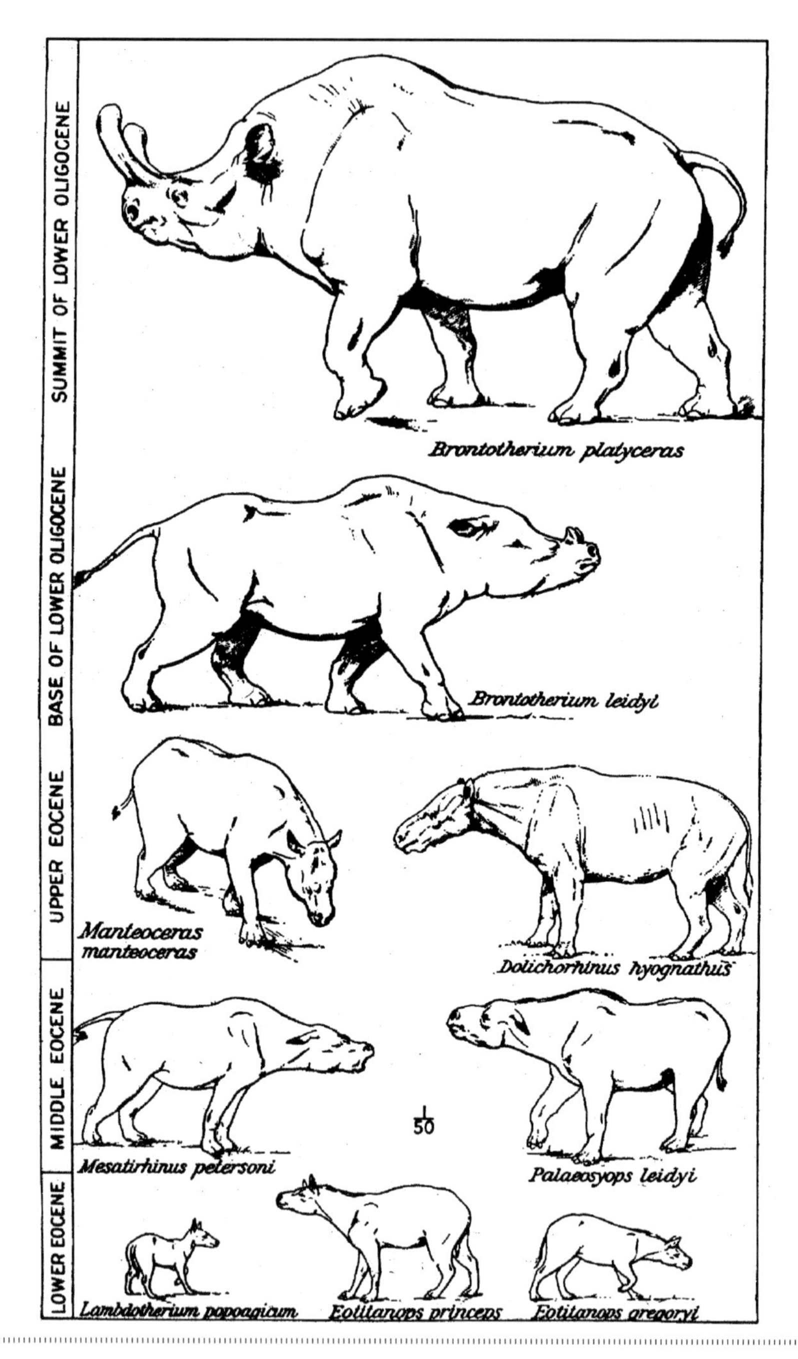

The supposed "orthogenetic" pattern of brontothere evolution, as suggested by Henry Fairfield Osborn (1929). Today, most of these brontothere taxa are considered invalid; their evolutionary history is branched and bushy, not linear, and they died out at the end of the Eocene.

4 | "Punk Eek" in the Badlands

As well as being lumps of patterned stone, fossils are also historical documents. History per se has had a bit of bad press recently.... There is a tension between the documentation of history (famously referred to as "one bloody thing after another") and the search for universal principles that are ahistoric and possibly timeless. After a period of doldrums, the bearers of historical tidings, the paleontologists, are making tentative movements toward the legendary High Table where, just visible through the clouds of incense (and rhetoric), the high priests of evolutionary theory smile benignly.

—SIMON CONWAY MORRIS, "EARLY METAZOAN RADIATIONS: WHAT THE FOSSIL RECORD CAN AND CANNOT TELL US"

A Paleontological Revolution

By the mid-twentieth century, paleontology had acquired the academic reputation as a musty, fuddy-duddy science. Up until the 1970s, most paleontology classes consisted of memorizing long lists of fossil names, their ranges in time, and the details of their anatomy. Classic textbooks of the early and middle twentieth century (such as Moore, Lalicker, and Fischer 1953) were almost entirely devoted to long detailed chapters with nearly every important taxon of fossil and almost no discussions of the theoretical principles behind paleontology. Most paleontology presentations at professional meetings were of the nature of "a new species of X" or "a new fauna from Y." Paleontology was notorious for its focus on small-scale detailed studies of such local problems. Relatively few paleontologists were focused on the "big picture" of patterns of evolution and faunal change through time. Stephen Jay Gould (1980b) called this mind-set the "idiographic" approach to paleontology: focusing on the trivial details and documenting lots of small-scale facts about the fossil record. There was little interest in what Gould (1980b) called the "nomothetic" approach: attempting to synthesize broader perspectives on the fossil record or deducing laws of nature from fossils. An infamous editorial in the journal *Nature* remarked that "scientists in general might be excused for assuming that most geologists are paleontologists and most paleontologists have staked out a square mile as their life's work" ("What Will Happen to Geology?" 1969:903).

Today, we look back at the 1960s as a time of revolutionary ferment, when the younger generation challenged society to change, and American society went through a huge struggle over civil rights, the Vietnam War, and the evils of a government that worked outside the law. This "question authority" attitude pervaded many other areas of life as well. The students who were educated and got their doctorates in paleontology in the 1960s may or may not have taken part in sit-ins and antiwar protest marches, but they were definitely revolutionaries. Many of them were more sophisticated than their graduate advisers in the latest ideas in evolutionary theory, mathematics, and especially the relatively new use of computers. They were eager to tackle questions and issues that were not on the menu of conventional descriptive paleontology. They asked broader questions about how the fossil record changed through time and what it could tell us about Earth and life history, not just documenting another specimen of species X. In the preface to his pioneering book *Models in Paleobiology* (1972), Tom Schopf described an encounter with a young Ph.D. student who could not decide on his thesis topic. He was considering describing a collection of fossils that his adviser had assembled. It had never occurred to the student to ask what problems he should be solving or what hypotheses he should be testing by using the fossil record.

The "Young Turks" of paleontology soon began to shake up the musty old profession, asking questions and introducing ideas that had seldom been the focus of paleontology in the past. The iconoclastic David Raup developed the first computerized models of fossil shapes and asked all sorts of unconventional questions about fossils that could be answered only by computers and mathematical models. Two young American Museum graduate students, Stephen Jay Gould and Niles Eldredge, were interested in biological issues, such as what speciation theory might tell us about paleontological species and how growth and development were important in understanding fossils. Steve Stanley got his degree at Yale working on the behavior of modern clams and how their shells functioned, with hardly a fossil in his thesis. And Schopf's volume collected many of these new ideas. In 1975, he and Ralph Johnson also founded the journal *Paleobiology*, which is still the premier forum for idea-based papers in paleontology some 34 years later. Indeed, the term *paleobiology* was coined to emphasize the biological, theoretical side of paleontology and to get away from the concept that all paleontology was boring descriptive "stamp collecting" with no ideas driving it.

In 1971, Raup and Stanley cowrote the revolutionary textbook *Principles of Paleontology*, which focused on the broad theoretical ideas and had *no* chapters describing fossils on a phylum-by-phylum basis. This book was my first text in paleontology when I was a college freshman in 1972, although I didn't realize how radical it was at the time. Ironically, I taught my first college classes in paleontology using Raup and Stanley's book, but after no one revised or updated it for many years, I decided that I had to write my own textbook on paleobiology.

By the time I was a young graduate student in the mid-1970s, the atmosphere at paleontology meetings was electric. There were intense debates over many topics, often polarized between the "old guard" trained before the mid-1960s who seemed to miss the point of the new thinking completely, and the "Young Turks" who were champions of the new approaches. The meetings sometimes got downright nasty as people shouted at each other across the room or resorted to insults and personal attacks—and then they'd rush home and pen a nasty article for the next journal issue, trashing their opponents. The scene was a far cry from the image of the cool, "objective" scientist as a neutral observer—but, then, that image is a myth fostered by the media and other nonscientific writers. Real scientists are human beings who can get passionate about important ideas and are willing to go to great lengths to defend them. It sounds a bit like politics or philosophy or other never-ending academic battles, but with an important difference: on most questions in science, there is a reality behind the ideas, and sooner or later enough data accumulate and enough hypotheses are tested that the bad ideas are abandoned, and only the good ideas survive.

The Eclipse of Evolutionary Paleontology

One of the consequences of this great flowering of paleobiology in the 1960s and 1970s was a renewed interest in evolutionary theory. In the early nineteenth century, when paleontology was in its infancy, the field was dominated by brilliant individuals such as Baron Georges Cuvier and Sir Richard Owen. They made important discoveries every time a new fossil came out of the ground. Cuvier was one of the first to recognize the principle of faunal change through the geologic record and to document the theologically unpopular fact of extinction. Owen not only named and described many of the most important early fossil vertebrates, but was the first to recognize and name the Dinosauria. When the idea of evolution came along in 1859, however, paleontology and the fossil record were not Charles Darwin's principal line of evidence. The fossil record was still relatively poor back then, with few fossils that showed the "insensibly graded transitions" that Darwin predicted should exist. Consequently, Darwin wrote two chapters in his book *On the Origin of Species* essentially apologizing for the incompleteness of the fossil record.

Just two years after his book was published, however, the first fossil of *Archaeopteryx* was discovered. Darwin's "bulldog," Thomas Henry Huxley, soon showed that *Archaeopteryx* was strong evidence that birds had evolved from dinosaurs. In the 1860s and 1870s, Vladimir Kowalewsky, Albert Gaudry, and Huxley all pointed to a series of fossils found in Europe that seemed to show the steps in the evolution of the horse. Then in 1876, Huxley came to the United States, where he saw O. C. Marsh's amazing collection of fossil horses at Yale. He immediately realized that horses had evolved primarily on this continent, with only occasional emigrations to Europe. During this heady time, examples of *Archaeopteryx*, horse evolution, and many other transitional fossils were discovered and described, further strengthening the paleontological evidence of evolution.

As the fossil record improved during the late nineteenth and early twentieth centuries, evolutionary theory itself went through a crisis. No legitimate scientist doubted the overwhelming evidence for the fact that life had evolved, but many evolutionists were not convinced that Darwin's mechanism of natural selection was the primary mechanism to explain evolution. The rebirth of genetics in 1900 brought with it a whole host of unconventional ideas, most of which did not regard natural selection as important. Other scientists revived the old notion of "inheritance of acquired characters" that is mistakenly labeled "Lamarckism" (even though Darwin himself advocated this idea). Those who argued for the primacy of natural selection, such as August Weissman, did numerous experiments that seemed to rule out any possibility of "Lamarckism." As a number of scholars have shown (e.g., Bowler 1985), this period was one of great diversity of opinion about the causes of evolution.

Paleontology reflected this divergence of opinion. Some paleontologists, such as Edward Drinker Cope and Alpheus Hyatt, argued for "Lamarckian" inheritance in their explanations of the fossil record. Others, such as Henry Fairfield Osborn, pushed their own idiosyncratic ideas even though they were not widely accepted by most paleontologists. Osborn was an advocate of the idea that lineages, such as his favorite group, the brontotheres (see the illustration that opens this chapter), evolved toward a goal of improvement (aristogenesis) and could evolve in a straight line without the constraint of selection (orthogenesis). Lineages might even evolve out of control to form deleterious features, such as the huge antlers of the "Irish elk" or the giant canines of the saber-tooth. According to Osborn, lineages reached a point where they had evolved into obsolescence (which he called "racial senescence"). These ideas were not credible then and are not so now, but because of Osborn's powerful position and influence, they were widely publicized. Still other paleontologists, such as Osborn's close friend and Princeton classmate William Berryman Scott, were agnostic as to the causes of evolution, even though they continued to document examples in the fossil record. Gould (1983) describes this research era as the period of "irrelevance," when paleontology had very little to contribute to evolutionary theory.

In the 1930s, a group of scientists worked out a method of mathematically simulating the process of evolution of gene frequencies through time. Known as *population genetics*, this research showed that even tiny amounts of natural selection were capable of producing major evolutionary changes. From this discovery, Darwinian natural selection enjoyed a rebirth called "Neo-Darwinism," which soon spread to conventional genetics with the work of Theodosius Dobzhansky in 1937 as well as to systematics and taxonomy with the work of Ernst Mayr in 1942. The first person to apply Neo-Darwinism to the fossil record was the most brilliant paleontologist of the twentieth century, George Gaylord Simpson. His book *Tempo and Mode in Evolution*—written in the late 1930s, but published in 1944 after delays due to the war—argued that Cope and Osborn's notions were outdated and wrong and that nothing in the fossil record was inconsistent with the newly developed principles of Neo-Darwinism. This argument eliminated the confusion and conflict of earlier generations, so that paleontology soon combined with systematics and genetics to form the Neo-Darwinian Synthesis. The synthesis reached its acme of influence by the centennial of the *Origin* in 1959, and there were almost no dissenters left.

As Gould (1983) and others argued, Simpson was almost too successful and placed paleontology in a role of subservience. He showed that the fossil record was consistent with Neo-Darwinism, but he gave the impression that the fossil record couldn't tell us anything about the evolutionary process that wasn't better studied with fruit flies or lab rats. Simpson himself protested that "experimental biology . . . may reveal what happens to a hundred rats in the course of ten years under fixed and simple conditions, but not what happened to a billion rats in the course of ten million years under the fluctuating conditions of earth history. Obviously the latter problem is more important" (1944:xxix).

Most paleontologists followed Simpson's lead and joined the great convergence of scientific opinion in the 1950s and 1960s, although there were a few dissenters. The great paleontologist Everett C. "Ole" Olson was a prominent force in paleontology during this time, primarily at the University of Chicago and then later in life at the University of California at Los Angeles (UCLA), where I got to know him very well. As he wrote in the volume of papers celebrating the 1959 centennial of the *Origin*:

> The statement is made, in effect, that those who do not agree with the synthetic theory do not understand evolution and are incapable of so doing, in most cases, because they think typologically. . . . Some avid proponents of the synthetic theory would appear to . . . eliminate as competent students of evolution, because of their inability to understand the theory, those who may disagree. . . . The situation proposes a frustrating dilemma for the sincere student who feels from his observations that there is more to evolution than can be studied, tested, and integrated under the synthetic theory, who is confident that real problems exist but also sees no way of making progress toward an understanding by means of the material that raise the questions in his mind. Few feel that the genetic-selection theory is invalid, but rather consider that there is much evidence that is not adequate. (Olson 1960:527–531)

Despite these protests, the Neo-Darwinian Synthesis was triumphant by the early 1960s, and the biology textbooks soon fell in line. Many of the leaders of the synthesis proudly proclaimed that the field had finally reached maturity and that the major principles of evolution had been solved. All that was needed was to fill in the small details with additional experiments and studies. Protests came from paleontologists such as Olson and others who felt that Neo-Darwinism could not explain many aspects of the fossil record. There was also dissent from many embryologists, who felt that their field had been ignored in the ascendancy of genetics. Evolution had been reduced to changes in gene frequencies through time, and whole organisms as well as their embryonic development were deemed irrelevant to evolution.

When people in a scientific field are saying, "We have reached the answer," it is a bad sign, not a good one. Science is always about criticizing, testing, and falsifying hypotheses, about never accepting anything as finally proven. Once scientists start talking about having the final answer, they are probably no longer critical enough of their own ideas and have become complacent. In the case of evolutionary theory, new developments *did* come along in the 1960s that challenged the Neo-Darwinian orthodoxy. They came from genetics, embryology, and many other fields, and argued that Darwinian natural selection in the form of changes in gene frequencies through time was not the only explanation for evolution. A detailed account of this challenge is presented in chapter 5 of my paleontology textbook *Bringing Fossils to Life* (2003) and in several of the works listed at the end of this chapter. Rather than repeat that entire discussion, I focus here on just one aspect of it: the challenge that arose from paleontology.

The Punctuated Equilibrium Challenge

As we have seen, many of the younger generation of paleontology students in the 1960s were ready to embrace new ideas and challenge the orthodoxies of the old guard. Some Young Turks were also much better versed in the new fields of genetics and systematics than were the earlier generation of paleontologists, who were trained before the synthesis became established as part of the standard curriculum and thus did not fully appreciate its implications. One of those new fields was known as *speciation theory*. Prior to the 1940s, most biologists did not worry about the mechanisms that formed species and assumed (as Darwin did) that changes in lineages through time were sufficient to account for the emergence of new species. During the 1940s, however, biological field studies and evidence from genetics had shown that the crucial component of speciation was genetic isolation. Large mainland populations, with their huge gene pools that had lots of gene flow (that is, interbreeding) seldom succeeded in producing new variants. By contrast, small isolated populations that were prevented from interbreeding with other groups and often had unusual gene frequencies were the apparent source of most new species observed in nature.

In the 1940s and 1950s, Ernst Mayr formally proposed and developed this idea in his allopatric speciation model. He argued that small, genetically isolated populations on the periphery of the main population (*allopatric* means "different homeland") are the key to the evolution of new species. Once the peripheral isolate population develops a new gene frequency, it can be repatriated alongside the ancestral sympatric population (*sympatric* means "same homeland"). If the genes have diverged enough so that the populations can no longer interbreed, they have become genetically isolated and distinct, and thus a new species exists.

These ideas were widely disseminated and accepted in the biology literature and textbooks by the 1950s, yet ironically most paleontologists of the 1950s and 1960s apparently did not see their implications for the fossil record. Most were still trying to document the notion that Darwin had first proposed: the fossil record should show "insensibly graded sequences" of organisms gradually changing through time as they evolved into new lineages. This notion of "phyletic gradualism" (figure 4.1A) was deeply ingrained in the thinking of many scientists, especially paleontologists, who were the masters of the long-term changes in the history of life. Some were even teaching about Mayr's allopatric speciation models in their classes, yet apparently not rethinking their own fossil record along those lines.

It took a younger generation of paleontologists to see the inherent contradiction between contemporary speciation theory and the classic notion of phyletic gradualism. In the 1960s, two doctoral students of the legendary paleontologist Norman Newell at the American Museum and Columbia University were Niles Eldredge and Stephen Jay Gould (figure 4.2). Both of these men were formidable intellects and soon became famous. When I first entered the Columbia–American Museum program as a new graduate student in 1976, there were legends about what it had been like a decade earlier when Gould and Eldredge were students in those hallowed halls. We heard stories from their contemporaries such as Jim Mellett about how Gould cranked out paper after paper with minimal effort. There was one legend that said that Gould had published a bunch of papers when Newell pushed him to finally finish his dissertation and get his degree, and that Gould then wrote his entire dissertation in a few days (or weeks or months, depending on which version of the story you hear). Whatever the truth behind these stories, these students' ideas and productivity were already legendary by my time, and every subsequent Columbia–American Museum grad student felt as if he or she were being measured by that impossible standard. Those were

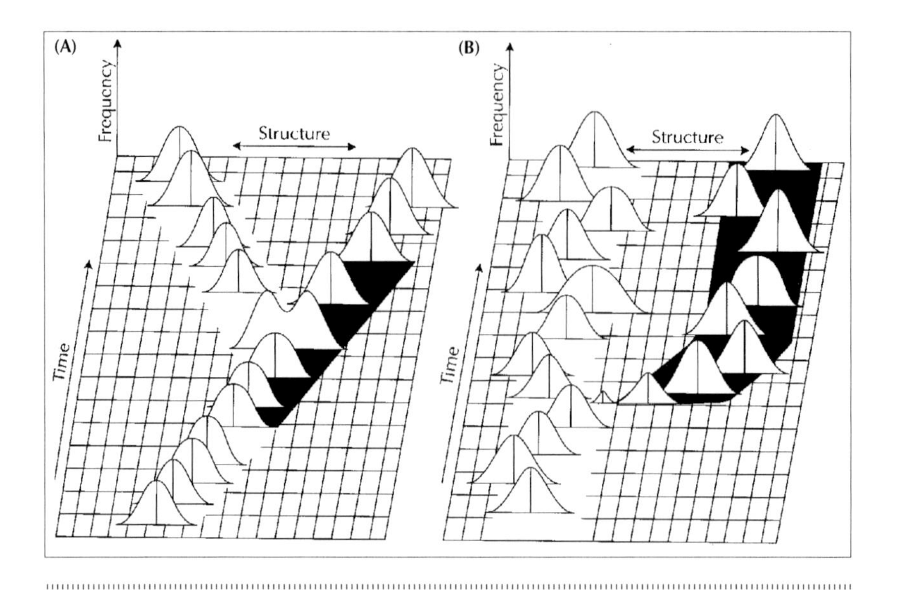

FIGURE 4.1 Patterns of evolution: (*A*) the classical Darwinian expectation of phyletic gradualism, according to which each population (represented by a histogram of its variability) gradually shifts through time (the vertical axis) to a completely different size or shape; (*B*) the pattern predicted by punctuated equilibrium, according to which most species arise through geologically rapid speciation events, followed by long-term stasis, where there is no net change in the histogram of population characters (although the characters may fluctuate around a mean value).

mighty big shoes to fill! (After I began taking classes at the American Museum, I was fortunate enough to be part of several grad student seminar classes taught by Niles and got to know Steve Gould very well when he came to visit New York.)

As Gould and Eldredge describe their academic history in several different books and articles, they both came to Columbia hoping to study evolution in the fossil record. Yet as Niles recalls in his 1985 book *Time Frames*, he found himself working on a project with these cool bug-eyed trilobites known as phacopids and was not finding a pattern of gradual evolution. Instead, each species appeared at a particular level in the fossil record and then remained unchanged through millions of years' worth of strata. Gould gives Eldredge credit for the insight (Eldredge 1971) that if you took the allopatric speciation model seriously, it would predict that species should arise in a normal biological time frame: a few years to a few hundred years at most. That's a geologic instant, the difference between one bedding plane and the next in strata that span millions of years. The allopatric speciation model also predicted that species should arise in small, peripherally isolated areas, so they were unlikely to be fossilized in the few places for which we have a good fossil record. Thus, the model accepted by biologists

FIGURE 4.2 (*A*) Niles Eldredge; (*B*) the author (*right*) with Stephen Jay Gould (*left*) and Michael Shermer, founder of the Skeptics Society, at Mount Wilson Observatory in 2001. ([*A*] photograph courtesy N. Eldredge; [*B*] photograph courtesy P. Linse and M. Shermer)

should give a fossil record where species appear suddenly without any gradual transition (punctuation) preserved and then persist for long periods of time without change (equilibrium).

This simple prediction led to the most cited paper in the history of paleontology, Eldredge and Gould's famous 1972 "punctuated equilibrium" paper. It was published in Tom Schopf's edited volume *Models in Paleobiology*, which included many other landmark papers advocating the new approaches to paleobiology. The paper is a masterpiece of writing and incisive thinking that poses a number of interesting issues. The first part is a general discourse on the philosophy of science, which points out that all scientists are products of their time and culture and tend to see what they expect to see. In this context, Darwin led paleontologists to expect phyletic gradualism, which they vainly tried to document for more than a century before the allopatric speciation model came along. Then Eldredge and Gould introduce the meat of the argument from the allopatric model, describe punctuated equilibrium, and give examples from their own research (phacopid trilobites from Eldredge, Bahamian land snails from Gould). Every time I teach a paleontology class, I always assign the original 1972 punctuated equilibrium paper as required reading and then lead a class discussion teasing it apart. Like fine wine, the paper gets better every time I reread it. I'm always amazed at what insights it contains, what future debates it triggered and foreshadowed, and how different students pick up different elements when they read it for the first time.

Although I was just a frosh taking my first paleontology class when the paper was published, by the time I got to grad school in 1976, it was already one of the most controversial papers ever published in paleontology and had triggered a flood of scientific literature attempting either to support it or to refute it. I vividly remember attending the Second North American Paleontological Convention in Lawrence, Kansas, in the summer of 1977, where we, McKenna's grad students, got our first exposure to the full spectrum of ideas and personalities in paleontology. Several sessions touched on the punctuated equilibrium debate, with all the major critics such as Art Boucot and Phil Gingerich getting in their licks, and Gould, Eldredge, Stanley, and many others defending it. The arguments were passionate on both sides, and it was an eye-opening experience to see such a major scientific debate played out on the stage in front of the entire profession of paleontologists.

The argument soon boiled down to just a few central points, which Gould and Eldredge (1977) nicely summarized on the fifth anniversary of the paper's release. The first major discovery was that stasis was much more prevalent in the fossil record than had been previously supposed. Many paleontologists came forward and pointed out that the geological literature was one vast monument to stasis, with relatively few cases where anyone had observed gradual evolution. If species did not appear suddenly in the fossil record and remain relatively unchanged, then biostratigraphy would never work—and yet almost two centuries of successful biostratigraphic correlations were evidence of just this kind of pattern. As Gould put it, stasis was the "dirty little secret" hidden in the paleontological closet. Most paleontologists were trained to focus on gradual evolution as the only pattern of interest and ignored stasis as "not evolutionary change" and therefore uninteresting, to be overlooked or minimized. Once Eldredge and Gould pointed out that stasis was equally important ("stasis is data" in Gould's words), paleontologists all over the world saw that stasis was the general pattern and that gradualism was rare.

The second major confusion was the issue of relative importance. Gould and Eldredge never made the absolutist claim that phyletic gradualism never occurs, only that the allopatric model would predict that gradualism would be much more rare than examples of stasis and punctuation. Indeed, this prediction has proven to be true for nearly every study since then. In some cases, there were examples of scientists pushing the case for gradualism in one particular lineage of fossils they studied. But when other scientists looked at the total data set, they discovered that the examples of gradualism were at best 1 to 2 percent of the total lineages in the fauna and that the entire rest of the fauna showed stasis and abrupt change. This situation was a classic case of biases putting blinders on some scientists, who then saw what they wanted to see and ignored the fact that the rest of the data were not consistent with their insistence on gradualism.

For example, Phil Gingerich of the University of Michigan was the principal advocate of gradualism during those years and kept pointing to examples of supposed gradual evolution in fossil mammals from the early Eocene of the Bighorn Basin in Wyoming. But when Gould and Eldredge (1977) analyzed his data in detail, they found that the supposed gradual sequences were evolving too slowly to be true biological gradualism and that you could break his "gradual" sequence into a series of rapid steps followed by stasis. More important, his gradual sequences were one or two exceptions in a fauna that was dominated by examples of stasis. In response, Gingerich made the absurd claim that stasis was "gradualism at zero rate," a revealing insight into his deeply ingrained biases.

Other debates common in those early days of the issue were similar in tone. Some scientists would point to a specific example of gradual evolution they had studied, but the usual reply to it was that the researchers had been biased in their data selection and that only a survey of the entire contemporary fauna would show the relative importance of gradualism versus punctuation. In other cases, there were claims of gradual evolution based on only three sequential samples, even though many successive levels with thousands of specimens and a rigorous statistical analysis are needed to show true gradualism.

Scientists often featured a single example of gradualism from a single section, but this means nothing if we don't look at the samples of the same fossil from different geographic regions. Otherwise, we might be looking at a lineage that is larger in body size to the north than to the south (a well-known geographic gradient called Bergmann's Rule). If this lineage had migrated through the sampled region over time, the local fossils would appear to evolve into larger or smaller forms. Only with a wide geographic spread (especially along a north–south gradient) can you rule out this issue. Gingerich's Bighorn Basin mammals were a good example of this problem. He focused on just a few lineages from the northern part of the basin, but his former student Dave Schankler (1981) showed that different patterns were apparent in the southern Bighorn Basin, just a few dozen kilometers away.

At each paleontology meeting in the 1970s and 1980s—for example, at each annual meeting of the Paleontological Society with the Geological Society of America (GSA) convention and at the Third North American Paleontological Convention in Boulder, Colorado, in 1987—the punctuated equilibrium versus gradualism debate still commanded entire sessions, and many people set out to test it again and again. A consensus was eventually reached. Stasis and punctuation appeared to be the norm for nearly all multicellular animals and plants, as would be predicted from the allopatric speciation model. Only a few possible cases of gradualism still remained that had not been shot down during this period of intense scrutiny and high scientific standards.

In microfossils, however, the story was clearly different. Many of them show true gradualism of shape change through hundreds of meters of deep-sea sediments, and stasis is rare. Most planktonic organisms are largely asexual clones with only limited sexual reproduction, though, so this gradualism is not surprising because they are not true sexual populations capable of genetic isolation as required by the allopatric speciation model.

The debate was less than a decade old when I was wrapping up my dissertation work in 1981. By this point, I had compiled an enormous database of nearly all the White River mammals. The purpose of this project was originally just to do Oppelian range-zone biostratigraphy and test the assumption of homotaxis, the idea that the same sequence of fossils in two or more sections of rock implies that they are contemporaneous (chapters 2 and 3). But one of the benefits of my huge data set was that I could also test the arguments about stasis and gradualism in the White River mammals. Gingerich's Bighorn Basin research was flawed by a narrow geographic spread and by his bias favoring the few gradualistic lineages in a mostly static fauna. By contrast, with the White River mammals, I had more than 160 well-dated, well-sampled lineages of mammals, so I could evaluate the relative frequency of gradualism versus stasis. I also had a wide geographic spread (from Montana and Saskatchewan to Texas, but mostly in the Dakotas, Nebraska, Wyoming, and Colorado). I had large samples of many species, with dozens at each level, and excellent stratigraphic data thanks to the careful work by Morris Skinner and the Frick field parties over several decades.

When I finally plunged in and plotted and analyzed my data carefully, it was clear that nearly every lineage showed stasis (figure 4.3), with one minor example of gradual size reduction (figure 4.4) in the little oreodont *Miniochoerus*. I could point to this data set and make the case for the prevalence of stasis without the criticism of bias in my sampling, as had plagued Gingerich. I vividly remember giving a presentation about this study for the first time at the GSA meeting in Indianapolis, Indiana, on Halloween afternoon in 1983. The crowd was packed for a very interesting session on a wide range of topics concerning evolution and extinction. I got up to give my talk a bit nervously because

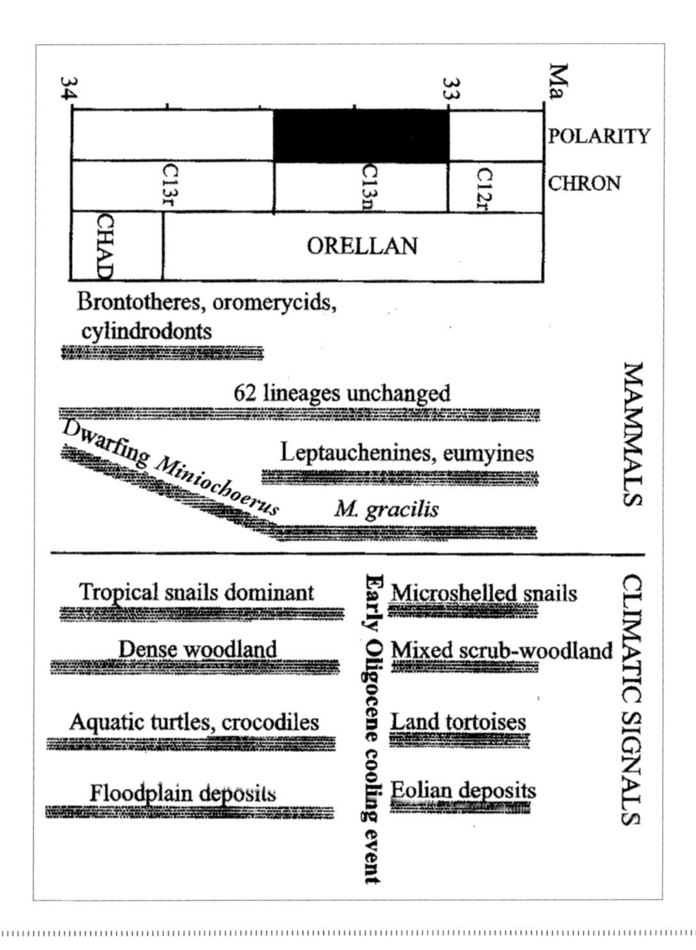

FIGURE 4.3 Summary of evolutionary patterns and evidence of climatic change in mammals across the Eocene–Oligocene transition in North America.

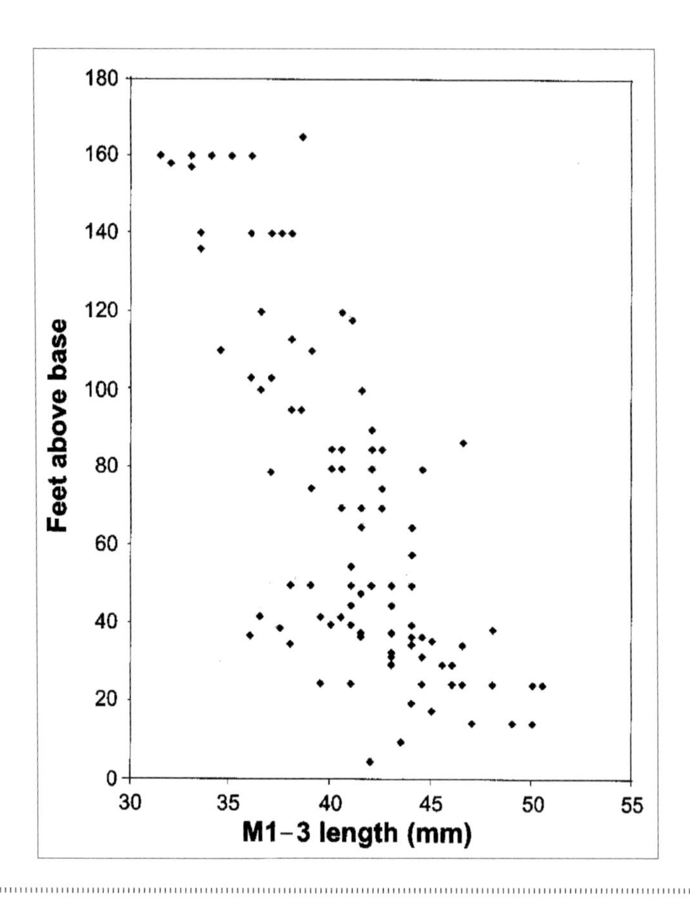

FIGURE 4.4 Gradual size decrease, or dwarfing, of the oreodont *Miniochoerus* during the late Eocene and early Oligocene as documented in the beds north of Lusk, Wyoming. The *x*-axis is a measure of size (as represented by the length of the three upper molars, M1–3), and the *y*-axis is height in section, a measure of time. The Eocene–Oligocene boundary falls about 25 meters (80 feet) above the base of the section.

many of the hard-core gradualists were still in the room and pretty hostile. But the talk was received well, and the most gratifying response of all was by Steve Gould himself, who got up and praised it effusively to everyone assembled. I was blushing with pride and gratitude when I got down off the podium after that testimonial!

More important, there in the crowd were my departmental colleague Dewey Moore and many students from Knox College, where I was teaching at the time. Just months earlier, the president of Knox College had closed down the geology department and eliminated my position (a year after I had arrived there). After

Dewey heard Gould's praise, he got Gould to write a wonderful letter refuting the Knox president's false claims that geology departments were unnecessary in small liberal arts colleges. Gould himself was an alumnus of little Antioch College in Yellow Springs, Ohio, and a disproportionate number of top geologists and paleontologists were also the products of the nurturing atmosphere of small liberal arts colleges. It was gratifying to see all the professional support we received as our department was under attack, as well as all the funding and pledges we received from loyal alums. We even had to call off a planned student strike and sit-in to trap the president in his office so that we wouldn't antagonize the Powers That Be any further.

But all this effort was in vain. When the trustees met in February 1984, they were forced to stand by their president, and we were cast aside. Dewey suffered tremendously through all this. At first, he escaped by taking a year of research leave on a Fulbright fellowship in Pakistan. As soon as the opportunity arose, he left Knox for the Illinois State Geological Survey, where he remained until retirement. After losing my job and interviewing for several new jobs, I was asked to stay on at Knox for one more year to help all the geology majors who were caught in the transition. The college had decided to keep the department open for one more year to let them graduate. Shortly after the president's actions, my dear friend and predecessor Larry DeMott, who had founded the department, died of a heart attack—a very sad event.

Meanwhile, I had my data and had given many talks showing the prevalence of gradualism in White River mammals, but I could not publish my findings yet because no one had done a detailed modern taxonomic revision of some important mammalian groups. In 1988 and 1989, I received a Guggenheim fellowship, which paid for a year off from teaching so I could go back to the American Museum and finish up my work on the evolutionary patterns and systematics of White River mammals. I eventually got Cambridge University Press to publish a volume edited by Bob Emry and myself (1996) that put most of this crucial data in print and in the same set of covers. Then Tim Heaton and I (Prothero and Heaton 1996) published a paper documenting the prevalence of stasis in nearly every White River lineage. By then, the fundamental data behind the study were complete and up to date, so the study was based on a firm foundation. I'm pleased to report that Gould gave it quite a bit of flattering coverage in his final work, *The Structure of Evolutionary Theory* (2002:861–865), his magnum opus on evolution, an immense tome of almost 1,500 pages published just weeks before he died. Even though I was never Steve's student in the formal sense (he was at Harvard while I was being trained at Columbia), he was always a good mentor and friend to me starting in my graduate student days. I was pleased that he regarded my research as important and that he featured some of my discoveries in several of his *Natural History* columns as well as in his final book.

Macroevolution?

When the original punctuated equilibrium paper came out in 1972, Eldredge and Gould briefly alluded to the possibility that the observed long-term stasis in species might have broader implications. But Steve Stanley at Johns Hopkins immediately saw another implication of this result: species appeared to be real entities that were resistant to change and not just an arbitrary assemblage of populations. He argued that certain distinct processes operated only on the species level, which in a pioneering 1975 paper he called "species selection." (In 1986, Elisabeth Vrba and Gould rechristened this concept "species sorting" to avoid the false implication that it is just an extension of natural selection.) Stanley followed his original paper with the landmark book *Macroevolution* (1979), which put the whole of evolution and the fossil record into this context of processes that operate at the level of species and not just individuals or populations. As Stanley, Gould, Vrba, Eldredge, and a number of other paleontologists argued, if species are distinct entities and not just collections of populations, then they are at a hierarchical level above that of individuals and populations. Just as individuals originate by the process of birth, species are formed by speciation; individuals die, but species go extinct.

Vrba and Gould (1986) and Norman Gilinsky (1986) pointed to a number of examples of animals that seem to demonstrate species sorting. Some lineages have properties that make it easy to speciate, whereas others do not and are relatively low in diversity throughout their history. For example, Thor Hansen (1978, 1980) argued that marine snails without planktonic larvae and therefore very little gene flow are prone to speciate more rapidly than those with planktonic larvae, and that prediction is confirmed in the fossil record. The larval condition is a property of the species, not of the individuals or populations, so it is driving this example of species sorting. In another example, two groups of gopherlike burrowing South American rodents exhibit very different properties that determine their speciation rates. The tuco-tucos (genus *Ctenomys*) are highly speciose because they have tiny home ranges and very limited gene flow. By contrast, the coruros (genus *Spalacopus*) have huge colonies with high rates of gene flow, so they are not very speciose.

Such examples have convinced many paleontologists, but the hard-core Neo-Darwinists still fail to see any difference between species sorting and natural selection. They come from a highly reductionist viewpoint, where bodies are just machines to spread more genes around (Richard Dawkins's "selfish gene" concept), and everything is reducible to simpler components. They are familiar with processes that operate on the relatively short timescales of years to hundreds of years, but seldom grasp the implications of long geologic timescales and millions of years of history documented in the fossil record. By contrast, those favoring the macroevolutionary concepts see the world as hierarchically ordered, in which the processes that affect populations are different than those that apply to individuals, and the processes that affect species are different than those that affect the population.

How might we go about testing these hypotheses? One approach is to show that species are real entities that must be treated as integrated wholes, not just a collection of organ systems or individuals that is constantly fine-tuned by natural selection. Here the paleontological record is excellent for just this sort of test. If the traditional Neo-Darwinian concept of evolution is true, and organisms are infinitely responsive to subtle changes in the environment (as suggested by the changes of the beaks of the Galapagos finches in response to climate change), then they should be constantly tracking changes in their environment.

However, paleontologists can track the changes in fossils and their environments for tens of millions of years. My own research in the White River Group proved to be an excellent test case. Much direct evidence indicates how much the environment changed from the upper Eocene Chadron Formation to the lower Oligocene Brule Formation. According to Greg Retallack (1983), the ancient soils suggest that the vegetation changed from thick forests to scrub brush (figure 4.5). According to work by Emmett Evanoff, myself, and Rob Lander (1992), the sedimentary rocks change from floodplain silts to dry dune deposits, and the land snails suggest a drying climate that changed from the ancient equivalent of the jungles of Nicaragua to the deserts of Baja California. J. Howard Hutchison (1982) showed that the late Eocene reptile fauna, dominated by crocodilians and aquatic turtles, was almost completely replaced by dry-land tortoises in the early Orellan. The oxygen isotopes of the tooth enamel of many of the mammals suggest a significant cooling and drying (Zanazzi et al. 2007). All of these findings are consistent with evidence from the marine realm (chapter 6) of a huge change in global climate and ocean temperatures and of the birth of the modern Antarctic ice cap at this time, when the world made a full transition from greenhouse climate to icehouse.

In our 1996 paper, Tim Heaton and I were able to show that among the White River mammals, there is almost no response to this huge environmental change during the early Oligocene (see figure 4.3). Out of 162 known lineages, most went right through the climatic change with no apparent change in size or anatomy. A handful of lineages became extinct or appeared at the climatic change horizon, and there was a dwarfing event in the oreodont *Miniochoerus* (see figure 4.4). But this response is so underwhelming in the face of the evidence for a huge climatic change that one is tempted to paraphrase Rhett Butler's final line in *Gone with the Wind*, "Frankly, my dear, the mammals don't give a damn."

I've presented this information in talks to dozens of different audiences, and so far I've found no simple explanation for why these mammals were so

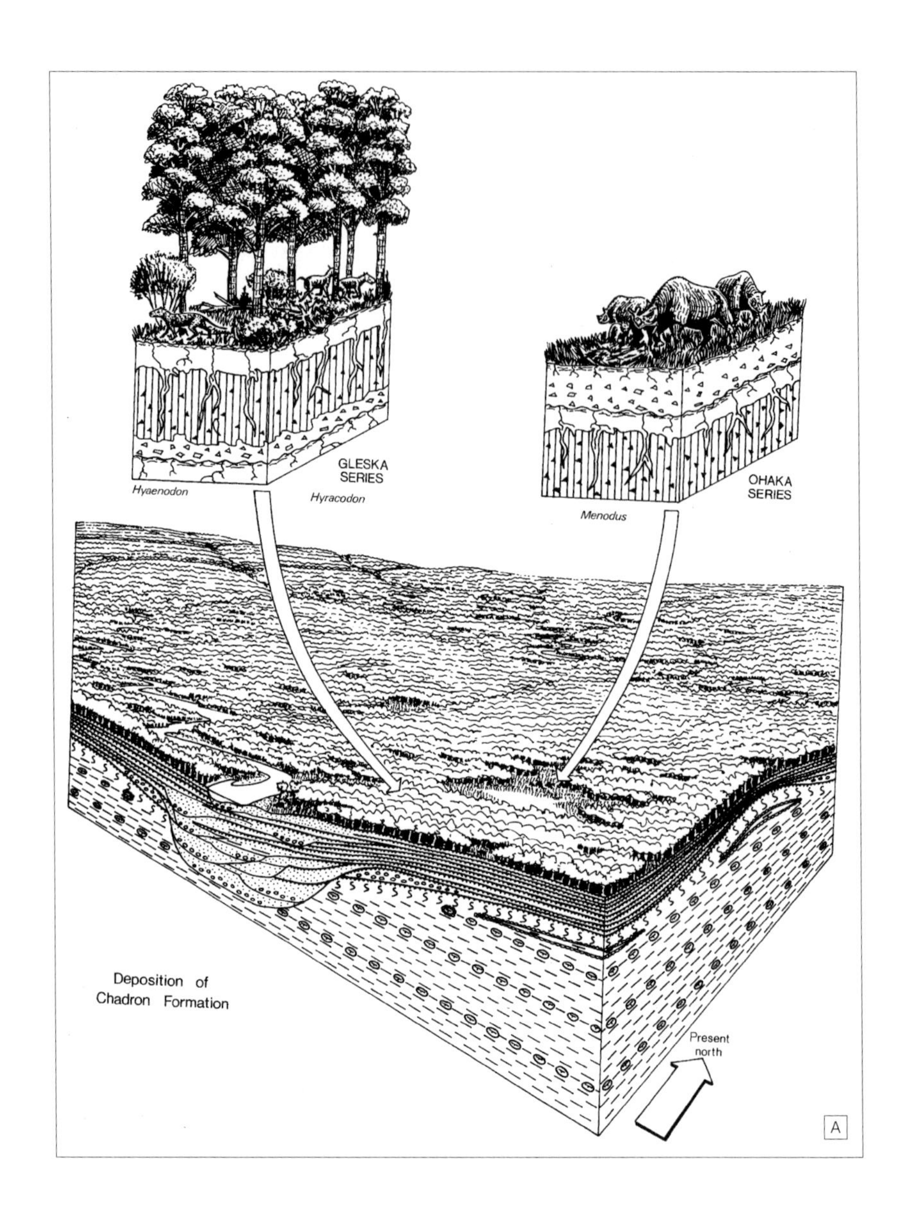

FIGURE 4.5 Transformation of the Badlands landscape according to evidence in ancient soils, as interpreted by Greg Retallack (1983): (*A*) the late Eocene was a time of dense forests with only limited open scrubby areas; (*B*) the late Oligocene, by contrast, was dominated by open scrublands with only limited tree cover near the riverbanks.

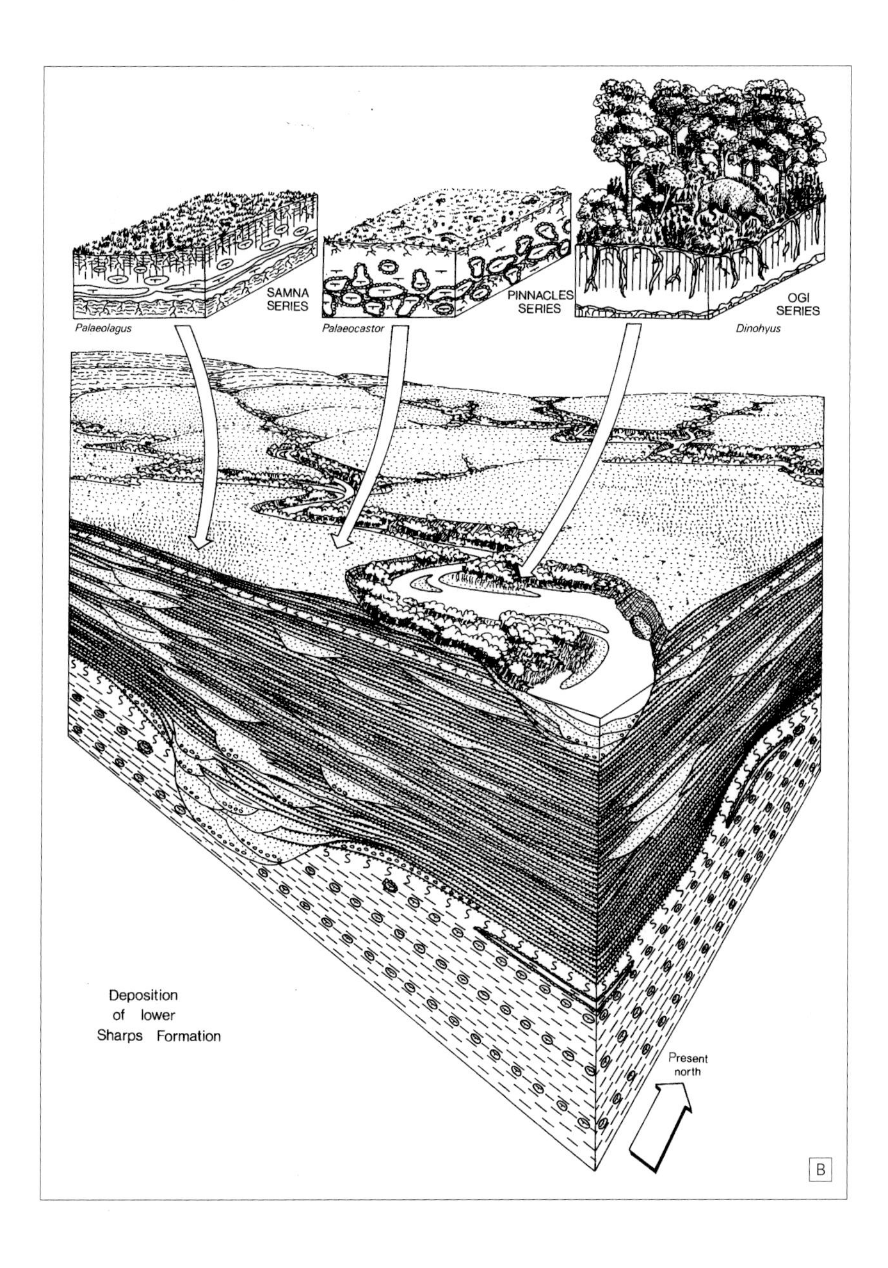

unresponsive to climatic change. Some people suggest that the mammals may have been generalists who could track a variety of different environments and vegetation types. More than 160 lineages of mammals, however, could not all be opportunists or generalists that could live on any kind of food; the survivors include the tapirs, whose teeth were highly specialized for eating leaves at a time when the forests almost vanished. In fact, Eric Dewar (2007) has looked at the scratches and pits on the surface of the teeth of nearly every group of mammals and finds that although the mammals show no changes in tooth shape, they were changing their diets noticeably. Clearly, their teeth did not evolve and adapt on the fine-scale "beak of the Galapagos finch" model, but retained their structural integrity and consistency of form despite notable changes in their function.

This observation intrigued me, so I began to reexamine the uncritical acceptance of the notion that fossil mammals always track environmental change. Vrba had published several papers (e.g., Vrba 1985) where she argued that fossil mammals of southern and eastern Africa were highly responsive to major climatic events, which drove them to extinction or speciation (the "turnover pulse" hypothesis). This idea was picked up and publicized uncritically in a number of books, especially because it seemed to correlate with major turnover events in the evolution of humans in eastern Africa. However, Kay Behrensmeyer and colleagues (1997) found no evidence of a "turnover pulse" in their much larger data set from northeastern Kenya near Lake Turkana, nor did Andrew Hill (1987) in his study of mammals from the Tugen Hills region. So the very foundation of the "turnover pulse" model was pretty shaky.

It occurred to me that our excellent database of North American fossil mammals and global climate change might be a good place to test the "turnover pulse" hypothesis. In a 1999 paper, I showed that for the four biggest independently documented periods of climatic change in the past 50 million years, the mammals either did not respond at all or showed much less speciation and extinction than they did at times for which there is no evidence of climatic change. One interval included the middle to late Eocene climate change at 37 million years ago (between the Duchesnean and Chadronian land mammal ages in North America), when turnover was merely at background level, despite evidence of floral change elsewhere in North America and of a significant climatic cooling event in the global oceans. The second was the Chadronian–Orellan transition just discussed. The third was the great expansion of modern grasslands at 7.5 million years ago, long after mammals with high-crowned grazing teeth appeared at 15 million years ago. In fact, there is almost no significant faunal change at 7.5 million years ago. The final example is the last 2 million years of ice ages, when climate changed dramatically, but speciation did not occur in response to the change. Instead, most Ice Age mammals simply migrated north and south in response to the movements of the ice sheets.

So are mammals just exceptions to the rule? Are most other organisms much more responsive to climate change, as predicted by the Neo-Darwinists? Not if the fossil record is to be believed. Some of my Caltech and Oxy students and I have just spent several years measuring all the common mammals and birds from the La Brea tar pits, not far from where I now live and teach. We find no evidence of any significant size or shape changes in all the vertebrates that are common enough in different-age tar pits. These animals include not only familiar mammals such as the saber-toothed cat, Ice Age lion, giant ground sloth, bison, horse, and camels, but even the common birds such as the condors, golden eagles, and turkeys. The tar pits span a period from 40,000 to 9,000 years ago, when climate changed dramatically from the previous interglacial through the peak glacial at 20,000 years ago and into the present interglacial at 10,000 years ago.

As more people look at the invertebrate fossil record, the same pattern is emerging. Steven Stanley and Xiangning Yang (1987) looked at the entire population of Atlantic Coast mollusks from the Miocene to Recent, but found little evidence of shape or size changes in response to the highly variable climates of the Miocene. Alan Cheetham (1986, 1987) found that the same was true of the bryozoans, or "moss animals." Jeremy Jackson and Alan Cheetham (1999) discovered little evidence for response to climatic change in a number of invertebrate groups. A whole new school of thought in paleontology called *coordinated stasis* argues that entire communities of animals are stable and unchanging for millions of years despite climatic changes. No matter where we look in the fossil record, organisms are much more stable in response to climate than the Galápagos finch model would suggest. They appear to have internal homeostatic mechanisms that resist small-scale change and respond only to a major event that might cause speciation or more likely extinction. We do not yet fully understand the genetic basis for this homeostasis, but it is real. For a long time, the common metaphor of a species was that it was rolling like a ball on the "adaptive landscape," with the ability to change easily in response to any slight external pressure and to change in any direction. Instead, our paleontological data suggest that a species is like a polyhedron. It has stable faces that keep the polyhedron in one position, and only after strong forces have tilted the polyhedron past the threshold does it quickly transition to another stable state.

The Neo-Darwinists naturally are not willing to accept this idea, and their critiques demonstrate that they don't "get it." For example, some (Charlesworth, Lande, and Slatkin 1982; Levinton 1983; Lande 1985) have dismissed the fact of stasis over millions of years of fossils as an example of "stabilizing selection." This is the type of selection documented in fruit flies that selects against the extreme variants of a distribution of organisms and keeps them centered on the average or mean type. But there is no way that stabilizing selection, a short-term process appropriate to biological scales of time, can apply to millions of years of time and strong environmental and climatic changes, for which we have well-documented evidence. The dry years in the Galápagos may cause some finches to evolve thicker beaks for eating the harder seeds, but this climate change may not be the cause of why they speciated and may not be relevant to what we see in the fossil record.

And so the question is still being debated, and the issue is not resolved. At the 1980 macroevolution conference in Chicago, the Neo-Darwinists faced the paleontologists on this issue for the first time, and the discussions were heated and passionate. Most still didn't see any reason to rethink Neo-Darwinism, yet some were beginning to realize that the fossil record was telling us something we could not see in fruit fly bottles or in Galápagos finches. Even Ernst Mayr himself (1992, 2001) later conceded that the degree of stasis in the fossil record was much greater than he expected and that biologists did not have a good explanation for it yet.

Clearly, a great deal is at stake. If the fossil record shows that the macroevolutionary patterns that we really care about are not the same as the microevolution of fruit flies and lab rats, but something operating on different scales with different processes, then almost a century's worth of research on those lab organisms is irrelevant to the problems we've been trying to solve since 1859. This is no trivial issue. We're talking thousands of published papers, hundreds of millions of dollars in grant money, plus positions, power, and prestige at stake. Someone raised in the conventional views of Neo-Darwinism will thus have a hard time understanding something that challenges their fundamental assumptions to the core, and they will resist and fail to appreciate what is going on until it is too late.

It is indeed an interesting time to be a paleontologist. From being labeled irrelevant in the early twentieth century and being placed in a subservient role in 1944, paleontology is now in the driver's seat and trying to reach the "High Table" where the "high priests" of evolution and genetics have long ruled unchallenged. Who knows where all this will end? Whatever the result, we're clearly making advances by challenging the accepted dogmas of the time, finding their faults, and, we hope, discovering something new and interesting. As Gould and Eldredge put it, "Why be a paleontologist if we are condemned only to verify what students of living organisms can propose directly?" (1977:149).

Further Reading

Eldredge, N. 1985. *Time Frames: The Rethinking of Darwinian Evolution and the Theory of Punctuated Equilibria*. New York: Simon and Schuster.

Eldredge, N. 1995. *Reinventing Darwin: The Great Debate at the High Table of Evolutionary Theory*. New York: Wiley.

Gilinsky, N. 1986. Species selection as a causal process. *Evolutionary Biology* 20:249–273.

Gould, S. J. 1980a. Is a new and more general theory of evolution emerging? *Paleobiology* 6:119–130.

Gould, S. J. 1980b. The promise of paleobiology as a nomothetic evolutionary discipline. *Paleobiology* 6:96–118.

Gould, S. J. 1982a. Darwinism and the expansion of evolutionary theory. *Science* 216:380–387.

Gould, S. J. 1982b. The meaning of punctuated equilibrium and its role in validating a hierarchical approach to macroevolution. In R. Milkman, ed., *Perspectives on Evolution*, 83–104. Sunderland, Mass.: Sinauer.

Gould, S. J. 1983. Irrelevance, submission, and partnership: The changing roles of paleontology in Darwin's three centennials, and a modest proposal for macroevolution. In D. S. Bendall, ed., *Evolution from Molecules to Men*, 347–366. Cambridge: Cambridge University Press.

Gould, S. J. 1992. Punctuated equilibrium in fact and theory. In A. Somit and S. A. Peterson, eds., *The Dynamics of Evolution: The Punctuated Equilibrium Debate in the Natural and Social Sciences*, 54–84. Ithaca, N.Y.: Cornell University Press.

Gould, S. J. 2002. *The Structure of Evolutionary Theory*. Cambridge, Mass.: Harvard University Press.

Gould, S. J., and N. Eldredge. 1977. Punctuated equilibria: The tempo and mode of evolution reconsidered. *Paleobiology* 3:115–151.

Prothero, D. R. 1992. Punctuated equilibrium at twenty: A paleontological perspective. *Skeptic* 1, no. 3:38–47.

Prothero, D. R. 1999. Does climatic change drive mammalian evolution? *GSA Today* 9, no. 9:1–5.

Prothero, D. R. 2003. *Bringing Fossils to Life: An Introduction to Paleobiology*. 2d ed. Boston: McGraw-Hill.

Prothero, D. R., and T. H. Heaton. 1996. Faunal stability during the early Oligocene climatic crash. *Palaeogeography, Palaeoclimatology, Palaeoecology* 127:239–256.

Somit, A., and S. A. Peterson, eds. *The Dynamics of Evolution: The Punctuated Equilibrium Debate in the Natural and Social Sciences*. Ithaca, N.Y.: Cornell University Press.

Stanley, S. M. 1979. *Macroevolution: Patterns and Process*. New York: Freeman.

Stanley, S. M. 1981. *The New Evolutionary Timetable: Fossils, Genes, and the Origin of Species*. New York: Basic Books.

Vrba, E. S., and S. J. Gould. 1986. The hierarchical expansion of sorting and selection: Sorting and selection cannot be equaled. *Paleobiology* 12:217–228.

The Cretaceous–Tertiary boundary at Gubbio, Italy, where the iridium anomaly was first discovered. The lower beds below the black boundary clay layer are uppermost Cretaceous in age, whereas those above the black layer are early Tertiary. An Italian lira coin is shown for scale. (Photograph courtesy A. Montanari)

5 | Death of the Dinosaurs

Mass extinction is box office, a darling of the popular press, the subject of cover stories and television documentaries, many books, even a rock song. . . . At the end of 1989, the Associated Press designated mass extinction as one of the "Top 10 Scientific Advances of the Decade." Everybody has weighed in, from the economist to *National Geographic*.

—DAVID RAUP, *EXTINCTION: BAD GENES OR BAD LUCK?*

The Age of Reptiles ended because it had gone on long enough and it was all a mistake in the first place.

—WILL CUPPY, *HOW TO BECOME EXTINCT*

Serendipity

Most people think that science is about planning your research carefully to achieve some specific goal. They are often not tolerant about "pure research" that doesn't have a specific conclusion in mind, but is focused on finding out general facts about nature, whether they have practical uses or not. Even the scientific funding agencies operate this way, rewarding research that is conventional and "more of the same," but seldom funding research that is a speculative gamble. Again and again, talking heads on television or in Congress ridicule "pure research" that doesn't have a specific practical goal or application. Narrow-minded and poorly educated people occasionally manage to interfere with the well-established scientific review process and shut down research they don't like, even though it was approved by established scientists.

The sad irony of the concept that "science must be practical and useful" is that most of the greatest discoveries in science happen by accident. More often than not, scientists who find a crucial new piece of evidence are not looking for it, but rather searching for something else and make their great discovery without planning to. The term *serendipity* was coined to describe this phenomenon. It comes from an old Persian tale, "Three Princes of Serendip," about princes who made discoveries unexpectedly. However, in the case of science, serendipity works most often when the researcher is prepared to see the implications of some new, unexpected development. As Louis Pasteur put it, "In the field of observation, chance favors only the prepared mind."

Examples of accidental discoveries in science are legion, especially in chemistry. Alfred Nobel accidentally mixed nitroglycerin and collodium (gun cotton) and discovered gelignite, the key ingredient for his development of TNT. Hans Von Pechmann accidentally discovered polyethylene in 1898. Silly Putty, Teflon, Superglue, Scotchgard, and Rayon were accidents, as was the discovery of the elements helium and iodine. Among drugs, penicillin, laughing gas, Minoxidil for hair loss, the Pill, and LSD were discovered by accident. Viagra was originally developed to treat blood pressure, not impotence. Most of the great discoveries in physics and astronomy were unexpected, including the planet Uranus, infrared radiation, superconductivity, electromagnetism, X rays, and many others. Two Bell lab engineers discovered the cosmic background radiation from the Big Bang when they were trying to eliminate the noise from their newly developed microwave antennas. Among practical inventions, inkjet printers, corn flakes, safety glass, Corningware, and the vulcanization of rubber were accidents. Percy Spencer accidentally came across the principle of microwave ovens while testing a magnetron for radar sets and finding that the candy bar in his lab coat pocket had melted.

Likewise, geologists often find things they are not looking for. In 1855, J. H. Pratt and George Airy were doing routine surveying for the British government in northern India. They noticed that the plumb line under the surveying tripod was not as gravitationally attracted to the Himalayas as they had expected and eventually discovered the evidence for the deep crustal roots of mountains like the Himalayas. The marine geologists who mapped the magnetic anomalies on the seafloor were not looking for the crucial evidence that proved plate tectonics, but were simply doing routine data collection of magnetic, bathymetric, and oceanographic data as their ships undertook regularly scheduled voyages of discovery. Maurice Ewing, the founder of Lamont-Doherty Geological Observatory (now Lamont-Doherty Earth Observatory) of Columbia University, had a standing order that each ship would take a deep-sea core at the end of the day, no matter where they were, and many of those cores turned out to have crucial evidence for the history of oceans, climates, and the evolution of life.

I can cite such examples for many more pages, but the point is clear: science is not always predictable, and scientific research cannot be restricted to straightforward results that are expected when the study begins. Shortsighted people such as the right-wing radio hosts and the politicians who ridicule pure research must not be allowed to destroy our scientific curiosity and creativity, or our scientific discoveries will come to an end. Isaac Asimov said, "The most exciting phrase to hear in science, the one that heralds new discoveries, is not 'Eureka!', but 'That's funny.'"

Out with a Bang . . . or a Whimper?

One particularly revealing example of serendipity in science occurred in Italy in 1978. Walter Alvarez, a young geologist who was working on the structural geology of Italy, was out mapping, measuring, and describing sections in the Apennine Mountains, the chain of mountains that form the "backbone" of Italy. (I knew Walter when I was a grad student at Lamont, and he was not yet famous, but just an ordinary structural geologist working in Italy.) Near the town of Gubbio, Italy, Walter came across a road-cut exposure (see the photograph that opens this chapter) that had thick limestones containing latest Cretaceous microfossils (as identified by Isabella Premoli-Silva), then another thick limestone with early Paleocene microfossils. In between them was a distinctive clay layer that marked the boundary between the Cretaceous and Tertiary, and thus one of the greatest mass-extinction horizons in Earth history. The Cretaceous–Tertiary (KT) boundary marked the extinction not only of the dinosaurs, but also of many marine organisms, including many types of plankton, marine reptiles, and the supremely successful ammonites, which looked something like the chambered nautilus and are related to squids and octopuses.

Walter knew the significance of the layer (the "prepared mind") but took a sample of the boundary clay thinking it would tell him something about how this great mass extinction had occurred. When he took it home to Berkeley and showed it to his father, Nobel Prize–winning physicist Luis Alvarez, they both puzzled over a way to unlock its secrets. Luis hit on the idea that rare elements that rain down on Earth from cosmic dust, such as iridium, might be useful. If, on the one hand, the sample showed a low level of iridium, then it would represent a relatively fast rate of sediment accumulation. If, on the other hand, it showed a high level, it would suggest a long accumulation of cosmic dust.

When they measured the sample, the iridium level was way off the charts and far too abundant to be the product of slow accumulation of cosmic dust. Luis eventually hit on a model that might explain this high level: the impact of an asteroid 10 kilometers (6 miles) in diameter, which had scattered debris rich in iridium all over the world and blocked out the sunlight with its dust cloud, causing mass extinction. Walter, Luis, and the two nuclear chemists who analyzed the samples at Berkeley, Frank Asaro and Helen Michel, finally wrote up their outrageous idea, and it was published in *Science* in 1980 (Alvarez, Alvarez, et al. 1980). It has since become one of the most famous and cited discoveries made in geology in the past 30 years.

Their original scenario went something like this: 65 million years ago a giant asteroid about 10 to 15 kilometers (6 to 9 miles) in diameter plummeted to earth.

It was traveling at cosmic speeds of 20 to 70 kilometers per second (45,000 to 156,000 miles per hour). Such an enormous mass traveling so fast packs an enormous amount of energy, approximately the equivalent of 100 million megatons of TNT, or more energy than all the nuclear weapons in the world at the peak of the Cold War. When the impact occurred, it would have generated a shock wave that should have leveled everything within 1,000 kilometers (620 miles) of the impact site, causing the world to burst into flames. In the imagined scenario, it would excavate a crater 15 to 20 kilometers (9 to 12 miles) deep and at least 170 kilometers (105 miles) wide. If the asteroid landed near or in the ocean, it would cause huge tsunamis. Approximately 100 cubic kilometers of rock and debris would blow from the crater and rise as high as 100 kilometers (62 miles) into the stratosphere. Most of this debris would fall back immediately, but some would generate a huge smoke and dust cloud that would blanket Earth for months and shut off all light to plants. In some scenarios, Earth would freeze over. In others, the impact would generate huge clouds of sulfuric acid that would be devastating for all of life.

After the 1980 paper announcing the iridium anomaly and the asteroid impact hypothesis, the scientific community was at first skeptical, and rightly so. Some wondered whether the iridium level was truly off scale and what the background level in the rest of the layers was like, so it took a great deal of work to establish

FIGURE 5.1 The Cretaceous–Tertiary boundary on land, here preserved in the Hell Creek Formation of North Dakota. The boundary itself is near the top of the section. (Photograph courtesy K. Johnson)

that the KT layer was indeed anomalous. Some scientists suggested that it might be due to the properties of marine clays in the Gubbio boundary layer. Clays are notorious for their ability to concentrate unusual trace elements, so this criticism was legitimate. When the same iridium anomaly was discovered in terrestrial rocks near Hell Creek, Montana (figure 5.1), however, the marine clay concentration hypothesis was ruled out. There was also one embarrassing gaffe in the original research: some samples were contaminated by the iridium from the platinum band of a technician's wedding ring. Even the high levels of the iridium in the boundary clay is less concentrated than your average piece of platinum-bearing jewelry.

By this point, many other scientists were jumping on the bandwagon of this exciting discovery. The extinction of the dinosaurs and the ammonites had long been one of the greatest unsolved puzzles of paleontology. Until 1978, no hard data existed that could test various wild speculations about supernovas, impacts, disease, sea-level retreat, and climatic change. The Alvarez hypothesis gave all sorts of people—such as geochemists, geophysicists, and planetary scientists with limited training in fossils and stratigraphy—an opportunity to plunge into the debate. As more and more people looked at KT boundary layers around the world, they found not only iridium, but also quartz grains that had undergone the effects of an enormous shock (previously known only from nuclear bomb craters), glassy blobs of crustal rock that had been melted and thrown into the atmosphere, and apparent tsunami deposits in the Gulf Coast and Caribbean. The impact bandwagon was rolling very fast among geochemists and planetary geologists, but not everyone bought into the idea.

The anti-impactors pointed out that there were other possible sources of iridium, such as deep-mantle-derived volcanoes like Kilauea. And the KT boundary was marked by one of the biggest mantle-derived flood lava eruptions in Earth history. Known as the Deccan traps, these eruptions are located in what is now western India and parts of eastern Pakistan (figure 5.2). They produced more than 10,000 cubic kilometers (2,400 cubic miles) of lava flows, with individual flows as thick as 150 meters (492 feet), and totaling at least 2,400 meters (7,874 feet) in thickness. Such enormous mantle-derived eruptions would have filled the atmosphere with thick clouds of ash-bearing iridium as well as with gases such as carbon dioxide that would have changed atmospheric and ocean chemistry.

Then yet another joker popped up in the deck. The latest Cretaceous was marked by a major drop in sea level, which drained the great epicontinental seas that once covered the High Plains from Hudson Bay to the Gulf of Mexico. The Cretaceous Interior Seaway had turned some places into a gigantic shallow ocean full of marine reptiles such as mosasaurs, plesiosaurs, gigantic turtles, huge fish longer than 7.0 meters (21 feet), gigantic clams 1.5 meters (5 feet)

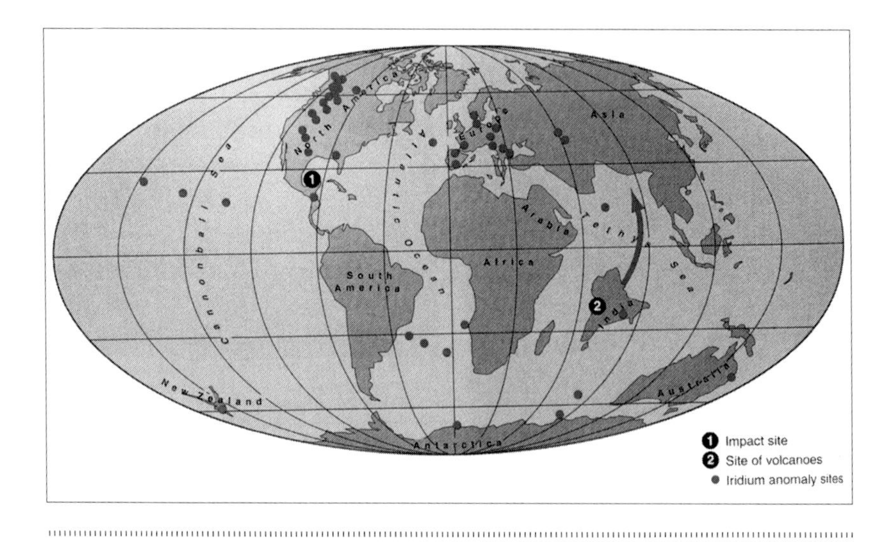

FIGURE 5.2 The location of the Chicxulub crater in the northern Yucatán (*1*), the Deccan traps in western India and Pakistan (*2*), and the sites of many of the iridium anomalies documented in nearly every land and marine sequence that crosses the Cretaceous–Tertiary boundary. (Redrawn from Prothero and Dott 2003)

across, and an incredible array of ammonites and other mollusks. This marine ecosystem was supported by a huge bloom of plankton whose skeletons are so numerous that they form the Cretaceous chalk that is widespread from the White Cliffs of Dover to western Kansas. Clearly, the drop in sea level and the drying of this seaway devastated shallow marine organisms and had an indirect effect on land life as well.

As the debate raged on through the 1980s, the biggest problem for the impact advocates was the absence of a crater of the right age. Then in 1990 planetary geologist Alan Hildebrand looked up some old oil company reports written by Glen Penfield in 1980. The oil companies had been drilling in the northern Yucatán at a place called Chicxulub (a Mayan word), where they had found a huge gravity anomaly that suggested some massive structure beneath the jungle. Some of their drill samples produced unusual rocks, such as shattered gypsum and possible impact debris. But they didn't find oil and weren't interested in the newly announced KT impact hypothesis, so their reports were quietly shelved. Hildebrand's "prepared mind," though, saw these findings as the "smoking gun" because all the strongest evidence for impact was around the Caribbean and Gulf of Mexico. Sure enough, when the impact advocates drilled much deeper under the jungle floor, they found an impact crater that had been filled in and buried, complete with abundant impact debris and argon-argon dates (analyzed by my friend Carl Swisher) that fit perfectly for the KT boundary.

By 1990, the debate was no longer whether the impact occurred—that was clearly established. Nor were the Deccan traps under any question—these immense eruptions occurred at just the right time. Nor was there any doubt about the great drop in sea level. With all these catastrophes happening in a short time window at the end of the Cretaceous, it was clearly a bad time to be alive on planet Earth. So the real issue was, Which of these three possible factors was most important? Did the impact all by itself cause the extinction, as the impact advocates argue? Or were the Deccan volcanics more important, possibly combined with the effects caused by sea-level change? The only way to answer these questions is to look at the fossils themselves.

What Do the Fossils Say?

If we're going to get beyond the name calling and wild speculation that has pervaded the research on the KT extinctions, we need to look closely at the evidence. Which organisms were affected? Which ones were not? Does the extinction show a gradual protracted pattern over the Late Cretaceous, or do most of the extinctions cluster right beneath the horizon bearing the impact debris and iridium? Fortunately, the intense interest in this extinction horizon has generated a great deal of research that answers just these kinds of questions. What do the fossils say?

First, let's look at the marine realm (figure 5.3). Five groups of plankton have readily fossilizable skeletons and make up the bulk of our marine record. The smallest are the tiny algae known as coccolithophorids, which are surrounded by a series of button-shaped calcareous plates only a few microns across, known as coccoliths. Another kind of golden brown algae are the diatoms, which secrete tiny plates that look like delicate Petri dishes made of silica. Yet a third group of algae, the dinoflagellates, have organic-walled shells propelled by tiny flagella. Feeding on these algae are amoebalike protistans that secrete a skeleton (unlike the amoebas familiar from your pond). These shells include the multiple bubblelike chambers of the calcareous shells of the foraminifera and the delicate siliceous "Christmas ornament" shells of the radiolaria.

Analysis of the plankton gave mixed results (MacLeod et al. 1997; Popsichal 1997). The tiny coccolithophorids did experience a severe extinction (as befitting a plant that requires light), but other algae, such as the dinoflagellates and diatoms, did not. Impact advocates argue that dinoflagellates and diatoms could

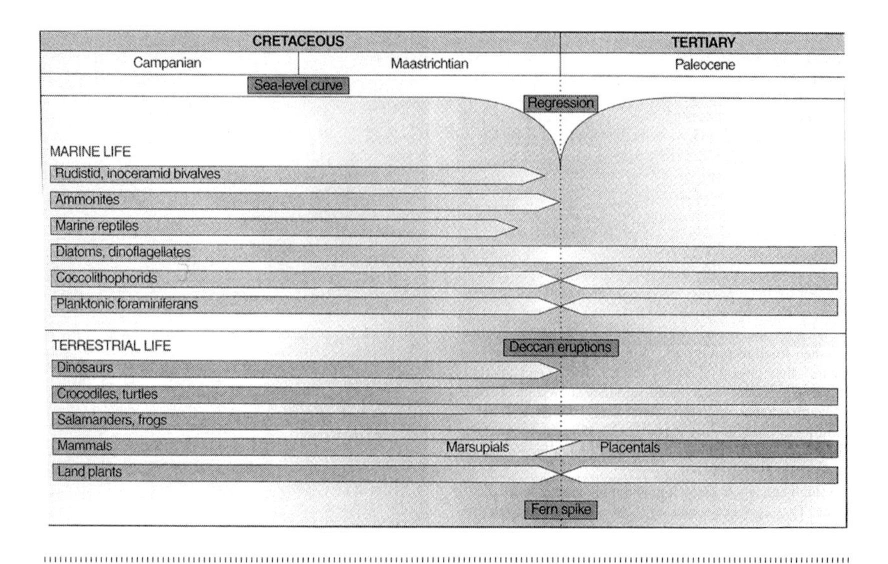

FIGURE 5.3 Summary of the pattern of extinctions near the Cretaceous–Tertiary boundary in marine and terrestrial organisms. (From Prothero and Dott 2003)

have survived in resting spores on the seafloor until the bad times ended, but this theory does not apply to the radiolaria, which also did not experience a significant extinction. In fact, the richness of radiolarians actually increased across the KT boundary, suggesting an improvement in oceanic circulation and productivity. Finally, the extinction in the foraminifera has been the most controversial over time. The bottom-dwelling benthic foraminifera show almost no extinction. Most micropaleontologists argue that the planktonic foraminifera suffered a significant and rapid extinction, but other researchers, such as Norman MacLeod and Gerta Keller (1995), argue that the extinction was protracted over 300,000 years around the KT boundary.

Up the food chain, the corals were in decline long before the KT event, and most "extinct" species of the latest Cretaceous reappeared in the Paleocene, so there is no clear evidence of a mass extinction of the coral reefs (Rosen and Turnsek 1989; Rosen 2000). Among the marine mollusks, there were the huge inoceramid clams and the cone-shaped rudistid clams that dominated the Cretaceous seafloor, but both groups were in decline long before the end of the Cretaceous and gone before the impact occurred (Kauffmann 1988; MacLeod 1994). The rest of the marine clams and snails show only a minor extinction, with about 35 percent of the snail species and 55 percent of the clam species dying out. Every study has shown that their extinction was gradual across the KT boundary (Hansen et al. 1987; Bryan and Jones 1989; Hansen, Farrell, and Upshaw 1989; Zinsmeister et al. 1989).

The most abundant group of mollusks to vanish were the ammonites, which had evolved and flourished after each of the previous mass extinctions, but did not survive the KT. Most studies have shown that they gradually disappeared through the Late Cretaceous, with only a few species surviving to witness the impact (Ward, Kennedy, and MacLeod 1991; Zinsmeister and Feldmann 1993). And their close relatives the nautiloids went right through the KT boundary with no extinction whatsoever and are still found in the South Pacific today. The squid-like belemnites, which once left thousands of shells shaped like large-caliber bullets, also declined throughout the later Cretaceous, with only one species around at the end to witness the impact (MacLeod et al. 1997).

There was an abrupt extinction in the brachiopods, or lamp shells (Surlyk and Johansen 1984), but not in their close relatives the bryozoans, or moss animals (Hakansson and Thomsen 1979). Nor was there much extinction in the echinoderms, including the crinoids (sea lilies), the sea stars, brittle stars, or the sea urchins and heart urchins (Birkelund and Hakansson 1982; Smith and Jeffrey 1998, 2000).

Finally, what about the marine vertebrates? More than 90 percent of the fish families survived, although their fossil record is too incomplete to determine how many species vanished right at the KT boundary. The marine reptiles, such as the dolphinlike ichthyosaurs and the long-necked plesiosaurs, were in decline long before the KT boundary and gone before the impact. The huge seagoing monitor lizards known as mosasaurs, however, were flourishing in the Late Cretaceous. With the huge sea-level retreat and the lack of shallow marine deposits at the KT boundary, it's hard to know whether they vanished abruptly at the impact horizon or not.

To summarize, the marine record sends a confusing mix of signals. Some extinctions, such as those of the planktonic foraminifera, the coccolithophorids, and possibly the brachiopods, are consistent with the idea that the impact was the dominant kill mechanism. Other animals—such as the benthic foraminifera, dinoflagellates, diatoms, radiolarians, and most of the clams, snails, nautiloids, echinoderms, and bryozoans—show relatively little effect at the KT boundary. And still other animals—such as the inoceramids, rudistids, belemnites, most ammonites, and the marine reptiles—were clearly in decline long before the KT impact and were not alive to witness it. No matter how you plead for the impact scenario, the facts show that it cannot have been as hellish as some advocates suggest, or there would have been a much more severe effect on the oceans.

What about the record on land? Once again, the biological record is a confusing mix of signals that cannot be explained by one simplistic mechanism. As J. David Archibald and Laurie Bryant show in their review of extinctions (Archibald and Bryant 1990; Archibald 1996), the dinosaurs are practically the only major victims of the KT asteroid impact. A number of scientists, however, have shown that the dinosaurs were already slowly declining through the Late Cretaceous, with only *Tyrannosaurus* and *Triceratops* vanishing at the end—and their youngest bones are 3 meters (10 feet) below the iridium anomaly in the Hell Creek beds in Montana (see figure 5.1). Out of 111 species documented in the Late Cretaceous, about 65 percent of the species survived the impact. In addition to the dinosaurs, there were significant losses in the sharks, marsupials, and lizards. Even more striking is which animals were *not* affected. The Late Cretaceous was rich in crocodilians, some of which were larger than the smaller dinosaurs, and they were relatively unaffected. Likewise, the turtles show no effects of the KT extinction, nor do the bony fish and amphibians. Although the fossil record of birds and insects is not as good, nearly all the Late Cretaceous lineages survived, suggesting that they were not decimated either (LaBandeira and Sepkoski 1993; Chiappe 1995). Douglas Robertson and colleagues (2004) tried to salvage the impact scenario by suggesting that all the survivors were able to burrow or seek shelter in lakes and rivers or in the ocean, but that explanation won't work with insects or birds, which survived, or with sharks, which didn't survive. No matter how one modifies the extreme effects of the "nuclear winter" version of the KT impact, it does not account for the majority of either extinctions or survivals at the KT boundary and must have been much less catastrophic than claimed. Instead, the slow decline of many of the species argues that extreme conditions, possibly triggered by the Deccan trap eruptions, may have been more important. Archibald (1996) looked at the known ecological characteristics of victims and survivors and showed that the sea-level drop actually explains the extinctions (especially of sharks) better than the more popular mechanisms. If anything, the impact was probably just the coup de grâce that finished off some of the survivors of this hellish time.

Another related scenario has been ruled out entirely. Several groups of scientists argued that the KT extinctions occurred because huge amounts of acid rain were produced when the impacting body hit the sulfur-rich basement rocks of the northern Yucatán Peninsula. But this idea can be dismissed because one of the groups to survive with almost no extinctions was the amphibians even though today the slight amounts of acid rain caused by human pollution are wiping out frog and salamander populations worldwide. If the huge acid rain bath scenario were true, there would not be an amphibian alive on the planet now (Weil 1984). Similarly, tropical bees cannot survive more than a few days if the climate becomes too cold or the flowers disappear, yet they did not vanish after the impact (Kosizek 2003).

Most paleontologists are still very skeptical because the paleontological evidence is not consistent with an impact that wiped out animals in the pattern that had been predicted. In addition, many geologists could think of other explanations. Through the 1980s and 1990s, Charles Officer, Charles Drake, Gerta Keller,

Norman MacLeod, and many others kept the debate going. I vividly remember the heated and often bitter scientific sessions at the annual GSA meeting when these two sides squared off against each other again and again. The arguments soon became very nasty and personal. The published literature was full of direct attacks on opponents' scientific credibility and competence, and the name-calling at meetings was even worse. Some people's careers were ruined because they were on the wrong side of the debate, and many others suffered enormously as the big boys battered it out. Luis Alvarez remarked: "I don't want to say bad things about paleontologists, but they're really not very good scientists. They're more like stamp collectors" (qtd. in Browne 1988). On the opposite side, Bob Bakker told a reporter: "The arrogance of these people is simply unbelievable. They know next to nothing about how real animals evolve, live, and become extinct. But, despite their ignorance, the geochemists feel that all you have to do is crank up some fancy machine and you've revolutionized science. The real reasons for the dinosaur extinctions have to do with temperature and sea level changes, the spread of diseases by migration and other complex events. In effect, they're saying this: we high-tech people have all the answers, and you paleontologists are just primitive rockhounds" (qtd. in Browne 1985).

There was also a deep cultural divide between the geochemists and planetary geologists on one side, who were used to fairly simple testable explanations based on data from big machines, and paleontologists on the other side, who were aware of the complexity of biological systems and especially the peculiar extinction pattern at the KT boundary. With such high stakes involved, each side in the debate had a great deal riding on sticking with its position and defending it to the end—not only scientific prestige, but also access to publications, grant money, lab space, tenure, and many other tangible and intangible rewards. No wonder the debate became so bitter and personal! It was and is a classic example of the sociology of science, where a deeply divisive scientific debate is largely propagated by the differences in scientific perspective as well as by stubbornness and unwillingness to consider opponents' arguments. Even though the arguments at the professional meetings have largely died down, the major players have not changed their minds and keep on publishing contrasting points of view some 29 years after the issue began (e.g., Keller 2005).

The geochemists and planetary geologists have claimed victory and dismissed all those who disagree with them as crackpots and fringe scientists. But that assessment cannot be valid when whole areas of the profession do not go along with the dominant view. In 1985, reporter Malcolm Browne of the *New York Times* took an informal poll of the paleontologists at the SVP meeting in Rapid City, South Dakota. (I was there, but somehow he missed me, probably because I was a young scientist with no publications on the topic.) The impact hypothesis was already five years old, but, according to the poll, the vast majority of scientists at that meeting found it unconvincing, and only 4 percent thought it was responsible for the KT extinctions. Even though some people have declared the impact hypothesis the winner and the matter settled, a number of books by well-respected paleontologists (Archibald 1996; Hallam and Wignall 1997; Dingus and Rowe 1998) beg to differ. The survey by MacLeod and his colleagues (1997) of a distinguished panel of 22 British paleontological specialists in nearly every group of marine fossils also came out against the impact scenario as the cause of marine extinctions. Then in 2004 another poll (Brysse 2004) was conducted of the SVP membership. Of those surveyed, 72 percent felt that the KT extinctions were caused by gradual processes followed by an impact. Only 20 percent felt that the impact was the sole cause. The other 8 percent had no opinion or questioned whether it was a mass extinction at all. Twenty-four years of arguments by the impact advocates apparently did not really change the opinions of the people who know the fossils the best.

Impacts Without Impact

During the early, heady days of the KT impact idea, scientists soon investigated the notion that impacts might cause other mass extinctions. They looked at the other major mass-extinction boundaries and tried to find iridium and other impact indicators. Some claimed there was evidence for impact at several of the other mass extinctions, although these reports proved to be premature. Nevertheless, I recall sitting in the audience at the International Geologic Congress in Washington, D.C., in 1989 and hearing Canadian paleontologist Digby McLaren (who once advocated a supernova as the cause of the Late Devonian extinctions) say that *all* mass extinctions were caused by impact, *whether there was evidence of impact in the fossil record or not*! I vividly remember the audience's stunned reaction at this blatantly untestable and unscientific statement. Ever the gadfly and provocateur, David Raup (1991) wrote that all extinctions (even normal background extinctions) might be caused by impacts. With such statements, why bother with data anymore? Impacts occurred, and extinctions occurred; therefore, all extinctions were caused by impacts.

Despite objections, the bandwagon for the generality of the impact-extinction hypothesis was rolling, and many scientists jumped on for the ride. As Keith Thomson put it, "With most subjects there is a silly season, usually of unpredictable duration and with an intensity correlated with the status of the acceptance of the new idea, [including] proposal of ideas even more far out than the original one" (1988:59). Paper after paper began to appear in *Nature* and *Science* claiming to find iridium or shocked quartz at many of the other extinction horizons. These papers were briefly peer reviewed, but were published quickly, often with great fanfare in the popular scientific media. Then a year or two later another group of scientists would try to reevaluate the data or replicate the results, and the "great discovery" would turn out to have been an illusion. The rebuttals were almost never published in these high-profile journals, however, nor did they get any publicity, so people only remember the first mention of the "discovery" and never realize that it was discredited.

In 1984, David Raup and Jack Sepkoski published a paper that claimed there was a 26-million-year periodicity in mass extinctions. Within weeks after it was published, numerous papers (based on the unpublished, unreviewed preprint of the article, according to Raup [1991]) came out trying to explain this "periodicity" in terms of periodic comet showers, the oscillation of the solar system through the galactic plane, an unknown Planet X, and even an undetected companion star to the Sun dubbed "Nemesis." Still others tried to explain the "periodicity" in terms of pulses of mantle volcanism. Within a few years, however, the idea itself was dead because the "periodicity" was not real, but rather an artifact of the methods that Raup and Sepkoski had used, along with a compilation of garbage data, bad statistics, fossil species that were not real, and bad timescale estimates (see the detailed review in Prothero 2003:chap. 6). In 1989, Sepkoski published a last-gasp defense of the idea, but Steve Stanley (1990) proposed a much simpler explanation for the apparent spacing of mass extinctions at 20 to 30 million years or longer. In a truly major mass extinction, life is decimated and reduced to a low-diversity world of a few "weedy" opportunistic survivors. It takes a full 15 to 20 million years for all the extinction-prone highly specialized species to reevolve. If some major crisis happens too soon after a major mass extinction, it has very limited effects because the world is inhabited by extinction-resistant survivors and has few or no vulnerable species.

When a bandwagon gets going, though, it is hard to stop. Planetary geologists and geochemists may not know much about biology or paleontology, but they are more than happy to jump in and take a few samples of a key boundary and make pronouncements. Only after time has passed and enough skeptical geologists with better training in stratigraphy and paleontology have reexamined the data do we really know if the initial discovery is real or not. For example, the "mother of all mass extinctions" was the Permo–Triassic event at 251 million years ago, which ended the Paleozoic era and may have wiped out 95 percent of all species on the planet. It is the greatest mass extinction to hit in the past 600 million years, so many geologists have tried to explain it. Various models based on global sea-level change, assembly of Pangea, and global cooling have been proposed and then shot down as better data are collected. In 2001, Luann Becker and colleagues argued that the extinction was caused by an impact and claimed to have discovered evidence at the Permo–Triassic boundary of some unusual forms of helium contained in fullerenes, which are the 60-carbon molecules also known as "buckyballs" after their geodesic structure and in honor of Buckminster Fuller. According to Becker and her coauthors (2001), these fullerenes with helium were products of a Permo–Triassic impact. These researchers then went on to identify Bedout Crater in Australia as the impact site (Becker et al. 2004). But Kenneth Farley and Sujoy Mukhopadhyay (2001) were not able to replicate Becker's "fullerenes" and found no unusual forms of helium that might indicate an impact. Several researchers (Glikson 2004; Koerberl et al. 2004; Renne et al. 2004; Wignall et al. 2004) quickly shot down the idea of the Bedout Crater, especially because it is the wrong age to have anything to do with the Permo–Triassic extinctions.

The same goes for all other mass extinctions that have been blamed on impact. The fifth biggest extinction in Earth history occurred between the Triassic and Jurassic periods of the Mesozoic, which eliminated 48 percent of the marine genera and was part of the process that shifted the dominant land vertebrates from the synapsids (formerly but incorrectly called the "mammal-like reptiles," even though they are not reptiles) to the early dinosaurs. In the 1980s, many geologists tried to tie this mass extinction to impacts as well. My friend Paul Olsen claimed there was evidence that the huge Manicouagan Crater in Quebec was the culprit (Olsen, Shubin, and Anders 1987; Olsen et al. 2002). This monstrous hole, which shows up as a huge ring on the satellite images of Quebec, is about 100 kilometers (62 miles) in diameter, not much smaller than Chicxulub. But recent redating of the crater debris puts its age at 214 million years ago, nowhere near the Triassic–Jurassic boundary at 201 million years or near the age of any other mass extinction (Palfy, Mortensen, and Carter 2000). Shocked quartz and iridium have also been claimed for this boundary, but further scrutiny has shown that their concentration was so small as to be unlikely to cause extinction (Hallam 1990, 2004; Hallam and Wignall 1997; Tanner, Lucas, and Chapman 2003). The third greatest extinction was in the Late Devonian (375 million years ago), when 75 percent of the marine species died out. Once again, impacts were blamed at first, and iridium anomalies were reported. However, closer scrutiny shows that these iridium anomalies are at the wrong time, and the evidence for impacts (if it is real) is not correlated with the several pulses of geochemical changes and extinctions in the Late Devonian (McGhee 1996).

The most striking example of scientists jumping on the impact bandwagon and getting ahead of their data concerned my own area of expertise, the Eocene–Oligocene transition. It is not one of the "Big Five" mass extinctions, but as we have already seen, it was a major event nonetheless. Raup and Sepkoski (1984, 1986) tried to fit it in their periodic-extinction model, although the "periodicity" predicted that the extinction should have occurred at 40 million years ago. In fact, there were two extinction pulses at 37 and 33 million years ago—nowhere even close! The Berkeley impact gang (Asaro et al. 1982; Alvarez, Asaro, et al. 1982) and several others (Ganapathy 1982; Glass, DuBois, and Ganapathy 1982) went looking for iridium. Sure enough, they found it *near* the Eocene–Oligocene boundary, so they crowed loudly in the scientific press about how they had "solved" the mystery of the Eocene–Oligocene extinctions. But paleontologists and stratigraphers who really knew the late Eocene (see the papers in Prothero and Berggren 1992 and in Prothero, Ivany, and Nesbitt 2003) dug in and did the hard detective work that the impact gang had completely neglected. The details that came to light showed that the impact was in the *middle* of the late Eocene at 35.5 million years ago, too late for the extinctions at 37 million years ago and too early for the extinctions at 33 million years ago. Except for a few species of radiolarians, there were *no* extinctions in any other group of organisms associated with these impact layers. "Close enough" may work in horseshoes and hand grenades, but not for precisely dated events such as those of the Eocene-Oligocene transition, where 1 to 2 million years of strata lie between each extinction horizon and the iridium anomaly.

In fact, this horizon has turned out to be an embarrassment for the impact advocates and a striking falsification of their simplistic view of the world. The sites of the middle to late Eocene impacts are now well known (Poag et al. 1992; Poag 1999; Poag, Mankinen, and Norris 2003). They include two big craters, one underneath Chesapeake Bay and the other at Toms Canyon on the Atlantic continental shelf, plus an additional impact site at Popigai in northern Siberia (figure 5.4). The Chesapeake Bay crater is huge, almost 100 kilometers (62 miles) in diameter. It is full of impact debris at the bottom, and since the impact occurred, it has filled with sediments from Chesapeake Bay over the past 35 million years. Popigai is only slightly smaller (90 kilometers [56 miles] in diameter). These craters are not as big as the 180-kilometer (112-mile) Chicxulub crater in Yucatán that is blamed for the KT extinctions, but they are close. Yet detailed studies of the crater and the strata formed in the aftermath show no evidence of extinction or of the global cooling predicted from the dust clouds, but actually a short global-warming event (Poag, Mankinen, and Norris 2003). Some last-ditch efforts (Poag 1999; Coccioni et al. 2000; Vonhof et al. 2000; Fawcett and Boslough 2002; Poag, Mankinen, and Norris 2003) have been made to salvage the impact-extinction scenario by blaming impact-induced climatic changes for the mass extinctions that happened 2.5 million years later, but these effects don't last in the atmosphere or oceans that long.

Actually, the lack of extinction for the second- and third-largest craters known after Chicxulub says something very different: impacts don't cause mass extinctions except in extraordinary cases. Wylie Poag (1997) replotted the "kill curve" that Raup (1991) had once fit to a single data point, the Chicxulub impact.

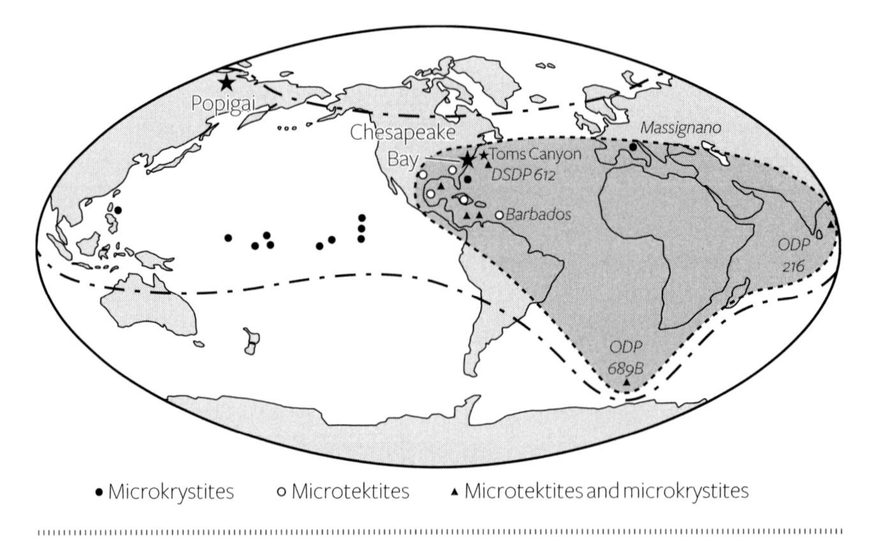

FIGURE 5.4 The location of the major late Eocene impact sites and their debris fields. DSDP = Deep-Sea Drilling Project; ODP = Oceanic Drilling Program. (After Poag 1999)

Raup's "impact kill curve" predicted that a crater of 100 kilometers (62 miles) in diameter should wipe out 60 percent of the species on the planet (figure 5.5). But when the 90- to 100-kilometer Chesapeake and Popigai craters are added in, the curve suddenly takes a very different shape. Apparently, *only* craters of nearly 200 kilometers (124 miles) in diameter are correlated with mass extinctions, and anything much smaller has little or no effect on life. Now even hard-core impact advocates such as Peter Ward (2007) are admitting that the impact bandwagon was premature and that impacts (except for the KT) have no effect on life. At the 2006 GSA meetings in Philadelphia, where the impact advocates had once reigned supreme, talk after talk was about the failure of impacts to account for extinctions and why the KT event is the sole exception. Some blamed extinction on the impact target. The KT Chicxulub crater was blasted into sulfur-rich gypsum bedrock, supposedly producing sulfuric acid rain, whereas all the other meteors hit granitic or gnessic continents or basaltic seafloor, which are not chemically reactive. As we saw earlier, however, this excuse is weak because the evidence of the negative effect of "sulfuric acid rain" is debunked by the fact that we still have frogs and salamanders. And as I already pointed out, the KT impact may not have had much of an effect after all if the paleontological data are to be believed.

In my scientific career, I've seen the profession go from puzzlement about mass extinctions before 1980 to the erection and dismantlement of the impact

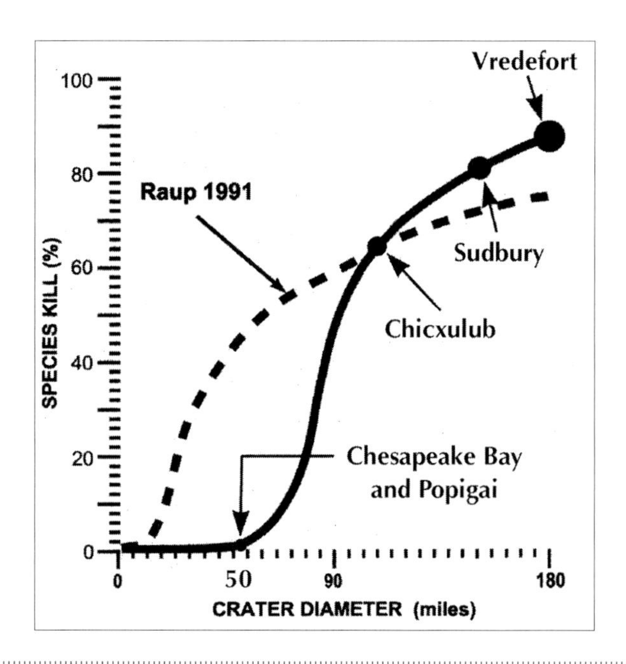

FIGURE 5.5 The "impact kill curve" as plotted by Wylie Poag (1997). In David Raup's (1991) original version (*dashed line*), the curve was fitted to only one point: the Cretaceous–Tertiary impact at Chicxulub. It predicted that a crater of 80 kilometers (50 miles) in diameter should produce a 40 percent extinction of species, a 145-kilometer (90-mile) crater should result in a 60 percent species extinction, and so on. But Wylie Poag (1997) refit the curve to the huge late Eocene craters at Chesapeake Bay and Popigai, and found that even craters of 97 kilometers (60 miles) in diameter produced almost no extinction. (From Poag, 1997, Fig. 4; courtesy SEPM)

bandwagon from 1980 to 2003, and now new ideas are coming along that may or may not better explain mass extinctions. The impact bandwagon beautifully demonstrates a salient fact about science: scientists are human and subject to social pressures and lured by attractive new ideas. But science is not like politics or philosophy, where one school of thought or idea can persist even after it has outlived its usefulness or in spite of much negative evidence. In science, we must measure our ideas against an external reality. Fads may come and go, and scientists may favor certain ideas for irrational reasons, but scientific hypotheses must stand the test of time and be corroborated by studies that may take years to finish. Our cherished ideas may ultimately turn out to be illusions, but as scientists, we cannot hang on to them and must move on. As Thomas Henry Huxley put it, this is "the great tragedy of Science—the slaying of a beautiful hypothesis by an ugly fact." Or as H. L. Mencken said, "For every problem, there is a solution that is simple, neat, and wrong."

Gassed to Death?

In November 2002, I found myself flying "across the pond" from Los Angeles to London and then running as fast as I could through Heathrow Airport customs and security to catch my Lufthansa flight to Berlin. I had been invited to speak at a major conference at the Humboldt Museum für Naturkunde in Berlin. The conference was titled "Mesozoic-Cenozoic Bioevents: Possible Links to Impacts and Other Causes." This huge old museum contains the five-story-tall *Brachiosaurus* skeleton from Tendaguru, Tanzania, featured in nearly every book on dinosaurs. Every time we stepped out of the conference room and looked out over the balcony, we could see its immense skull towering above us from our balcony view on the second floor. It was a striking reminder of the topic of extinctions that was the focus of the conference. The museum contains many other paleontological treasures, including the best-known and most complete of the 11 known specimens of *Archaeopteryx* (the "Berlin specimen") as well as amazing dinosaurs and marine reptiles collected in the late nineteenth and early twentieth centuries, when Germany led the rest of the world in the field of paleontological exploration. This museum amazingly survived the bombing and shelling of Berlin during World War II, even though most other German museums were destroyed and lost countless irreplaceable paleontological treasures. During my first visit to Berlin as a high schooler on a two-month tour of Europe in 1971, I had no chance to visit the museum because it was on the other side of the Berlin Wall. We had to reach Berlin by train through then Communist East Germany and saw the sights of West Berlin. My high school tour group was allowed to pass through Checkpoint Charlie and briefly drive around some of East Berlin on a tour bus, mostly to witness the grim desolation of the Communist side of the city. In 2002, the Berlin Wall and Checkpoint Charlie were gone, replaced by Cold War museums and new streets, and Germany was unified and economically strong and thriving again.

The organizers of the conference included my grad school buddy and frequent coauthor Dave Lazarus, who was now a curator of micropaleontology at the museum, and several meteorite specialists who were big fans of the impact model. The impactors presented their talks the first day of the meeting, but for the rest of the conference the paleontologists got up and presented the newer data that made the impact model obsolete. Many of the leading figures in the mass-extinction debates were there, so the questions were pointed and intense. I was practically the only speaker focusing on the Eocene–Oligocene transition, but I made a strong case that the impact model, the mantle-volcanism model, and many others were inadequate to explain the transition. This was first time I saw the impact-extinction hypothesis go into disfavor and other models start to replace it.

The impact-extinction model may be on the decline, but that does not stop scientists from looking for a common pattern among the great mass extinctions. The most popular alternative to the impact model has been the mantle-volcanism model, which argues that during times of gigantic flood basalt eruptions, huge amounts of gases are released from the mantle that may cause extreme climate change and even runaway global warming (Rampino and Stothers 1988; Courtillot 1999). This model receives support from the fact that three of the Big Five mass extinctions are definitely associated with huge mantle-derived volcanic eruptions. I have already mentioned the Deccan traps, which erupted just before the end of the Cretaceous. The Triassic–Jurassic extinctions correlate with huge eruptions from the Central Atlantic Magmatic Province, which took place when the North Atlantic began to rip open as Pangea broke apart.

The Permo–Triassic extinction was coincident with huge eruptions known as the Siberian traps, the largest such eruptions known in Earth history. These flows were up to 6,500 meters (7,100 yards) thick in 11 discrete eruptive sequences and covered a total of about 7 million square kilometers (2.7 million square miles), an area equivalent to that of the continental United States (Erwin 2006). The ages of these flows have been recently redated at 252.2 and 251.1 million years old, exactly the same as the dates from the Permian–Triassic boundary in China. It would be remarkable if eruptions so immense did *not* have an effect on late Permian life.

The problem with this attractive model is that none of the other major mass extinctions correlate with major volcanic episodes. In addition, many huge eruptions occurred that did *not* cause mass extinctions. No major volcanic episodes have been shown to correspond to the complex Late Devonian extinction or with the Late Ordovician extinction. Conversely, no mass extinctions are associated with the Columbia River flood basalts (15 to 16 million years old) that cover most of eastern Oregon and Washington. The eruption of the North Atlantic Tertiary province is dated between 61 and 56 million years ago, but it is not associated with any significant extinction either. Several researchers (Rampino and Stothers 1988; Courtillot 1999; Courtillot and Renne 2003) attempted to attribute the Eocene–Oligocene extinctions to flood basalts in Ethiopia and Yemen. Unfortunately, the latest dates on these lavas are between 29.5 and 31 million years, or during the middle Oligocene, which makes them several million years too young to have had anything to do with the extinctions at 37 and 33 million years ago.

But there may be an indirect link between some of these eruptions and another possible killer: atmospheric gases. The trendiest new idea is that some of the major mass extinctions (especially the Permo–Triassic, Triassic–Jurassic, and possibly the Paleocene–Eocene extinctions mentioned in chapter 1) were caused by unusually high concentrations of carbon dioxide and low concentrations of oxygen in the atmosphere (Ward 2006, 2007). The evidence for this explanation comes from the models of atmospheric gases developed by Bob Berner at Yale (Berner et al. 2003), which show unusually high levels of carbon dioxide and low oxygen at these three extinction horizons, as well as at the Late Devonian, Late Ordovician, and Late Cambrian extinctions (figure 5.6). Further evidence for this model comes from the geochemistry of seafloor sediments, which show unusual enrichment in light carbon that might have come from eruptions such as flood basalts. In fact, the current model for the Permo–Triassic extinction involves oversaturation of the oceans with excess carbon dioxide (called *hypercapnia*), which is fatal to most marine life. Likewise, the high carbon dioxide and low oxygen levels of the Permo–Triassic and Triassic–Jurassic boundaries would have been extremely stressful for most land animals and particularly favored those with more efficient respiration, such as the early dinosaurs (Ward 2006).

At the GSA meeting held in Denver in October 2007, this trendy new idea was promoted at the Pardee Symposium and given maximum publicity. It certainly seems plausible given what we know now and worthy of further examination. Nevertheless, it is not the be-all or end-all of explanations of mass extinction. For example, there is no evidence that the Eocene–Oligocene extinctions suffered from elevated carbon dioxide or low oxygen, either on the Berner curves or in any other data source. In fact, the transition from greenhouse climate to icehouse climate suggests that carbon dioxide was declining, not spiking, in abundance at the mass-extinction horizons. Some people have tested specific predictions made by Peter Ward (2006) and found that they contradict or even falsify his model (Holtz 2007), so it is premature either to declare it as the final explanation for mass extinctions or to rule it out either. As occurred with its predecessors, collecting enough data to evaluate it and either accept or reject it will take years. More likely, we may find that it fits certain extinction events, but not others.

Eocene–Oligocene extinctions are notably the joker in the deck that always seems to falsify the generalized attempts to explain mass extinctions, from the Raup and Sepkoski periodicity model to periodic volcanism to the impact scenario and now the "gas attack" model. No matter what we do to massage and tweak the data for these extinctions, they stubbornly refuse to go along with any single pattern deduced from any other mass extinction. The Eocene–Oligocene extinctions stand out, resistant to simple explanation, as the persistent falsifier that ruins any attempt to give a unified explanation to mass extinction. I didn't realize it at the time I started research on the Eocene–Oligocene, but I could not have asked for a more interesting and important interval to build my career on. It's just another example of serendipity.

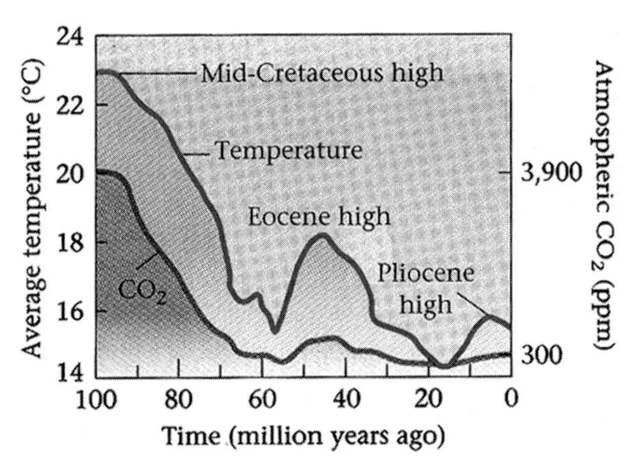

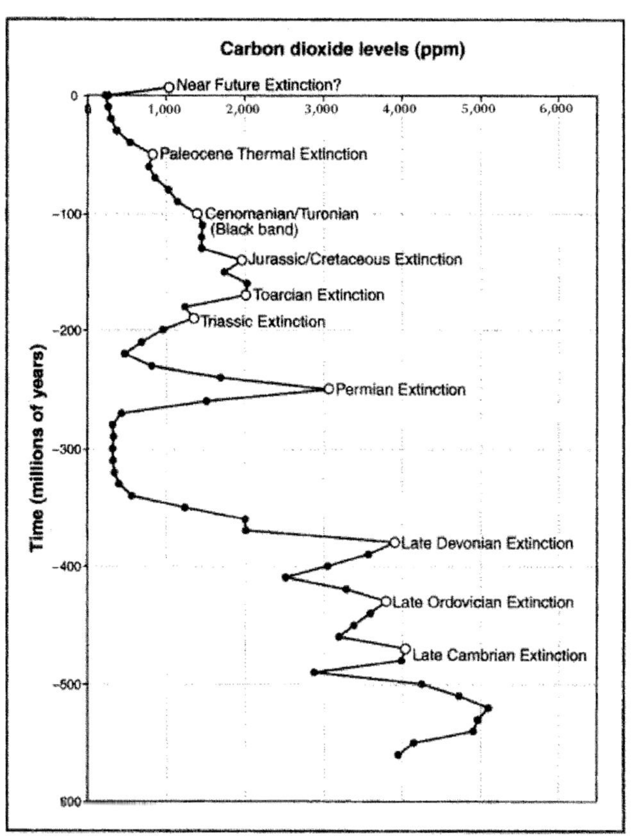

FIGURE 5.6 (*A*) Trends in carbon dioxide (CO_2) and global temperature through the late Mesozoic and early Cenozoic, showing the early Eocene warming event and the Oligocene cooling; (*B*) carbon dioxide levels through the past 500 million years, according to Robert Berner and his colleagues (2003). (Modified from Ward 2007)

Further Reading

Archibald, J. D. 1996. *Dinosaur Extinction and the End of an Era: What the Fossils Say*. New York: Columbia University Press.

Courtillot, V. 1999. *Evolutionary Catastrophes: The Science of Mass Extinction*. Cambridge: Cambridge University Press.

Erwin, D. 2006. *Extinction: How Life on Earth Nearly Ended 250 Million Years Ago*. Princeton, N.J.: Princeton University Press.

Glen, W., ed. 1994. *Mass-Extinction Debates: How Science Works in a Crisis*. Stanford, Calif.: Stanford University Press.

Hallam, A., and P. B. Wignall. 1997. *Mass Extinctions and Their Aftermath*. Oxford: Oxford University Press.

MacLeod, N., and G. Keller, eds. 1995. *Cretaceous–Tertiary Mass Extinctions: Biotic and Environmental Changes*. New York: Norton.

Officer, C., and J. Page. 1996. *The Great Dinosaur Extinction Controversy*. New York: Addison-Wesley.

Poag, C. W. 1997. Roadblocks on the kill curve: Testing the Raup hypothesis. *Palaios* 12: 582–590.

Poag, C. W. 1999. *Chesapeake Invader*. Princeton, N.J.: Princeton University Press.

Prothero, D. R. 2003. *Bringing Fossils to Life: An Introduction to Paleobiology*. 2d ed. Boston: McGraw-Hill.

Prothero, D. R. 2005. Did impacts, volcanic eruptions, or climatic change affect mammalian evolution? *Palaeogeography, Palaeoclimatology, Palaeoecology* 214:283–294.

Raup, D. M. 1986. *The Nemesis Affair: A Story of the Death of the Dinosaurs and the Ways of Science*. New York: Norton.

Raup, D. M. 1991. *Extinction: Bad Genes or Bad Luck?* New York: Norton.

Ward, P. 2006. *Into Thin Air: Dinosaurs, Birds, and the Earth's Ancient Atmosphere*. New York: Joseph Henry Press.

Ward, P. 2007. *Under a Green Sky: Global Warming, the Mass Extinctions of the Past, and What They Can Tell Us About Our Future*. Washington, D.C.: Smithsonian Books.

Collecting paleomagnetic samples from the middle Eocene Coaledo Formation in the Coos Bay region in southwestern Oregon in the summer of 1999. My field crew consisted of Clio Bitboul (*left*) and Elizabeth Sanger (*right*). (Photograph by the author)

6 | Marine World

If I have seen further it is by standing on the shoulders of giants.

—ISAAC NEWTON, LETTER TO ROBERT HOOKE

Pied Piper of the Mudflats

August 12, 1998. My student crew and I are wading at low tide along the north shore of the Olympic Peninsula between the tiny towns of Twin Rivers and Pysht, Washington (see the photograph that opens this chapter and figure 6.1). The slime and muck are unbelievable, and the place stinks of rotting algae and seaweed as well. Each time we take a step, our foot sinks in. Each time we pull out a foot with great effort, it makes a loud sucking sound, and we are up to our knees in slime. My student Linda Donohoo does this once, and her Teva sandal doesn't come out of the muck, so she has to probe around the deep mud to find it. In addition, the weather is overcast and drizzly, so we are soaking wet on top and muddy on the bottom. Not exactly a pleasant way to spend a day—and we have to hike several miles through these conditions from rocky headland to headland covered with slippery, slimy boulders, through bay after bay filled with slime. We have to sample every decent outcrop of the upper Oligocene Pysht Formation as quickly as we can because this low tide lasts only a few hours. If we don't reach our access point to the road before the tide comes in, then we'll have to scramble up steep cliffs and hack our way through a jungle to find our vehicles. Otherwise, we'll be drowned as the tide comes in and the waves pound the shoreline once again.

Our stalwart guide through this miserable trip is Casey Burns, one of the best amateur paleontologists in the state of Washington. Casey reminds one of Santa Claus with his jolly face, belly, and bushy white beard. Like Santa Claus, he wears long black boots, but they are for wading in mud and water, not Christmas snow.

FIGURE 6.1 Collecting paleomagnetic samples on the low-tide mudflats on the northern shore of the Olympic Peninsula, August 12, 1998. In the center with the pack and walking stick is Casey Burns. Casey is talking to Clio Bitboul. Stooping to the right is Liz Nesbitt of the University of Washington Burke Museum. Behind her is Linda Donohoo, and behind Casey is Elana Goer. (Photograph by the author)

He also carries a long walking stick he decorated with his own carvings. It helps him balance when he's scrambling over the rocks. And his pervasive jolliness and high spirits help buoy up our morale as the conditions get more and more difficult. Any time our spirits are about to crater, he tells another funny story, and we push on.

Actually, a better nickname for Casey might be the "Pied Piper of Washington." Casey is leading us because he works along this coastline for fun. His primary occupation is making amazing hand-carved wooden flutes from his cozy home in drizzly Kingston, Washington, across Puget Sound from Seattle. Every few months, he takes his flutes to trade shows and Renaissance fairs, where they are considered works of art. But whenever there is a really low tide on the calendar, Casey and other legendary amateur paleontologists such as Jim Goedert take time off work and wade along the Oregon or Washington coasts, looking for fossils. They wake up before dawn to catch the lowest tides of all, just like the early-morning fishermen, but their quarry is long dead, turned to stone, and wouldn't taste good at the dinner table. They walk and wade for hours when the tidal exposures are at their best and often find amazing specimens, including some of the earliest toothed whales, baleen whales with teeth, and other spectacular marine

FIGURE 6.2 Doug Emlong proudly showing off one of his discoveries. (Photograph courtesy Oregon Department of Geological and Mineral Industries)

mammals. They prefer low tides during the winter, when there is no coating of algae and rotting slime to cover the rocks, but the downside of working in the winter is that the weather is often bitterly cold and rainy. The fossils are sometimes contained in huge concretions with only a few bones showing, requiring Casey or Jim to pack as much as 100 pounds on their backs through the mud and slippery rocks to reach the nearest road. At other times, they spend an entire day scouring the coastline until the tide comes in, but find little or nothing. Nevertheless, their efforts have produced some of the most important fossils that demonstrate the origins of the modern groups of whales and many other marine mammals.

One of the most legendary of these amateur fossil collectors was the late Doug Emlong (figure 6.2). He was a fanatic about collecting for fun, even when it conflicted with his job responsibilities. Starting in 1956 at age 14, he would work the Oregon and Washington coasts at every low tide, with incredible tenacity and dedication through horrible conditions. But the hard work paid off. He found and described *Aetiocetus*, one of the oldest known toothed baleen whales, as well as *Behemotops*, the oldest known member of the order Desmostylia (a peculiar group of extinct coastal mammals that looked like beach hippos). *Behemotops* was found in this very coastal stretch of the Pysht Formation where Casey, my crew, and I are struggling in the mud (see figure 6.1). This specimen helped prove the relationships of desmostylians to elephants and solved a long-standing paleontological puzzle. Emlong also found the oldest known complete skeleton of the transitional fossil later named in his honor, *Enaliarctos emlongi*, demonstrating the ancestry of seals and sea lions from bears. He also solved a great mystery, finding the best skull of the peculiar "beach bear" *Kolponomos*, which showed that it was just a highly modified bear adapted for coastal feeding. The list of his famous finds goes on and on. Even though he was an amateur with very limited training in paleontology, he had a true nose for fossils and could find things when no one else could. He tried building a homemade museum of his discoveries in Depoe Bay, Oregon, but late in life he struggled with depression and mental illness. Then in 1980, at age 38, he vanished one day. He had apparently committed suicide in a giant collapsed sea cavern called the Devil's Punchbowl near Otter Rock, Oregon. Fortunately, Emlong had developed a good relationship with Dr. Clayton Ray of the Smithsonian, who made sure his collection ended up in a scientific repository, and many different scientists have worked on his amazing discoveries ever since. Quite a few of these finds are named *emlongi* in his memory.

Such amateur collectors are often the backbone of the profession in some places. There are a few professional paleontologists at the universities and museums in western Oregon and Washington, but their time is entirely consumed by teaching classes and attending committee meetings and running museums. They typically have very little spare time for fossils, and most of their nonteaching time is focused on research and publication of fossils that have been collected but never studied. Amateurs who can spare the time from their jobs often are the best collectors, anyway, because they are dedicated and work long hours when necessary—and some, such as Goedert, Burns, and Emlong have an amazing talent for spotting fossils as well. Without them, our understanding of marine mammal evolution would be nowhere near as complete as it now is.

Return to the Pacific Coast

I found myself wading along the Washington coast in 1998 because of a shift in focus in my research career. I had been working the Badlands and other Eocene–Oligocene mammal-bearing rocks of the Great Plains and Rocky Mountains since my first field season in 1979. After completing all of my planned grant-funded work in the Rockies and Plains in 1986, 1987, 1988, and 1991, there weren't many other important Eocene–Oligocene localities left to do magnetic stratigraphy on, certainly nothing that would impress the NSF or the Petroleum Research Fund, which had generously funded my research for years.

When I moved back to California in 1985, I soon realized that my own backyard along the Pacific Coast offered great opportunities for research. This new focus actually started in 1982, when I was attending a meeting of the American Association of Petroleum Geologists (AAPG). Back in those days, the AAPG meeting was peculiar from my perspective as an academic geologist focused on pure research because it centered entirely on petroleum geology and most of the talks were thus very narrow and limited to the specific problems of a particular oil field or exploration techniques. The slides shown were beautiful because oil company geologists had plenty of money to pay professional artists to work for them, but compared to the GSA meeting each year, the climate for ideas at the AAPG meeting was very confined and sterile. The GSA was a marketplace of amazing new ideas, even though the academic geologists showed crude homemade slides. Now, with Powerpoint, the playing field for making pretty slides has been leveled. Even my undergrads can do amazing slides, so AAPG no longer has the edge in presentation quality.

I was at the 1982 AAPG meeting to give a presentation on my magnetic stratigraphy of the White River Group and its implications for the global timescale. While I was there, I sat in on a talk by the legendary John Armentrout, a top-flight geologist at Mobil Oil Corporation. John was not only famous for the excellence of his stratigraphy and exploration work, but also well known for his piercing gaze, intense focus on his research, and ramrod-straight posture. He had lightning-quick recall of almost any aspect of geology that he'd learned and an amazing depth of intelligence. He was also a forceful speaker who could rivet an audience when he wanted to. He had gotten his Ph.D. in 1973 at the University of Washington in Seattle working on the marine Eocene–Oligocene rocks of the Olympics and their fossil mollusks. He had revised the molluscan biostratigraphy of the region and had revolutionized the stratigraphy as well. Back in the early 1970s, the best job opportunities for a young stratigrapher-paleontologist were in the oil business, and John had risen quickly through the ranks at Mobil.

At this particular meeting, John stood in front of the assembled oil geologists to describe some of his past work in the Pacific Northwest and ended his talk with a plea. The marine rocks of the Olympics had been studied long ago by legendary geologists who by the 1940s had worked out the formations' molluscan biostratigraphy. After that period, there was relatively little restudy of them by professional paleontologists except for Armentrout himself, whose research efforts belonged to Mobil. Yet they held enormous promise for further research, especially with new methods of dating and other important techniques. Armentrout's clarion call to the audience was to find a collaborator who might be interested in working on these promising exposures—improve their dating and correlation, study their isotopes, and exploit their underused potential.

After John's talk ended, I intercepted him as he left the lecture hall and told him of my interest. We were soon making plans for my first field season as a professional Ph.D.-bearing paleontologist to sample his most important sequence, the 3,000 meters (10,000 feet) of strata of the Oligocene Lincoln Creek Formation along the Canyon River on the south side of the Olympic Peninsula. Once I had my first Petroleum Research Fund grant to do fieldwork in the summer of 1983, I put these rocks on my agenda—the same field summer that I described in the beginning of chapter 3, when my crew and I survived being stranded behind Sheep Mountain Table and collecting record numbers of samples in a single day in the John Day beds of central Oregon.

As soon as my crew and I left the John Day region, we headed north to the Olympic Peninsula and set up camp in a true jungle. The Olympics are legendary for their 3.6 meters (140 inches, almost 12 feet) of yearly rainfall on the western slopes, forming one of the few temperate rain forests in the world. We were camped in the southwestern Olympics, but the vegetation was just as dense, with huge Sitka spruces overshadowing impregnable thickets of ferns and vines with incredible numbers of thorns. We set up our tent near an old logging road, but had to wait a day when the rains came and the river level rose several feet in a few hours. Once we had dry weather, we dressed in shorts, light jackets, wading shoes, and a backpack for our samples and tools. Unlike our previous collecting on steep desert cliff in the Badlands, this kind of sampling would require wading across the Canyon River again and again, sampling little tiny patches of outcrop on the cutbanks of each bend in the river (figure 6.3). The rest of the bedrock was covered by a dense jungle, so there were no exposures except those scoured clean by the river.

On July 8, 1983, we plunged into the river for the first time and discovered its unique challenges. The vegetation was so dense that we couldn't see anything outside the narrow river valley around us, so I learned to navigate by the shapes of each bend in the river. John had provided excellent detailed maps, so I knew precisely where we were with respect to his fossil localities. Nowadays, a geologist will carry a portable geographical positioning system (GPS) unit to find his or her position on a map (as I now do), but such devices did not exist in 1983. With the forest towering overhead, however, even today I typically cannot get a GPS fix on enough satellites when I work in these narrow confined places. I'm eternally grateful to my undergrad geology professors at Riverside who taught me all about how to navigate and find myself on a map under any set of conditions. Unlike some of my students, I navigate by careful compass work and map reading, so I'm seldom lost, and I'm never a slave to a new technology like GPS that doesn't always work.

FIGURE 6.3 The Canyon River section. Sampling along the Canyon River on the southern Olympic Peninsula in July 1983. My crew—(*left to right*) Allison Kozak, Annie Walton, and Rob Lander—struggles to take paleomagnetic samples on the slippery, narrow ledges before wading into the icy river water yet again. (Photograph by the author)

We found that the first challenge was getting *to* the river. There were only a few places where the logging road approached one of the river bends, so we had to find such an access point and work our way through the jungle until we could drop into the cold water. Nearly every plant in that part of the world has thorns or stiff, sharp branches. Some of my crew wore jeans to protect against the thorns, but others (like me) just bushwhacked through the jungle in shorts. Boy, was I grateful for the cold clear water on my scratches and cuts when I finally stepped into the river! Once we were on the riverbed, we had to navigate from one creekside exposure to the next, usually by zigzagging across the river as we walked. In most places, the river was very low and knee-deep for us, as it typically is after a relatively dry summer in the Olympics. However, we soon learned how to avoid the slick rocks with coatings of slimy green algae, which would make us tumble and land on our butts. We also had to avoid those deeper blue-green regions of the river because they would turn out to be scoured by the current and deeper than our waists. More than once, one of us would misjudge the river and go from water that was only knee-deep to a sudden full-body plunge. This water was cold glacial melt from the Olympic Mountains to the north, so it was chilly! If the clouds and cold winds came along and we got wet, we'd find ourselves shivering and doing all we could to prevent hypothermia before we could hike all the way back to the truck.

Nevertheless, we plugged on like this day after day, taking unscheduled breaks whenever a rainstorm would interrupt us. We would seek shelter in my tiny Eureka Timberline tent until the river level dropped again. One evening, a huge storm came up, and the river almost rose high enough to flood our tent site, but we slept through the entire thing blissfully unaware and only realized the next morning how close we had come to being washed away. Finally, we had collected every important outcrop along the river spanning more than 3,000 meters (10,000 feet) of thickness of the Lincoln Creek Formation, and the field summer was almost done. We had enjoyed much of our stay in cool, comfortable Washington State, but we were grateful to return to the hot, dry Big Badlands in South Dakota for our final visit to Peanut Peak (chapter 3) before we headed back home. Rob Lander and I trimmed and analyzed the samples at the South Dakota School of Mines in Rapid City by late August, and the results were excellent.

I quickly wrote up the results and plotted them, so that John Armentrout and I would be coauthors on our magnetostratigraphic work on the Lincoln Creek Formation. Two years later the paper was finally published (figure 6.4) in the widely read journal *Geology* (Prothero and Armentrout 1985). John had chosen the section well because the Canyon River section of the Lincoln Creek Formation was the thickest and most complete exposure of upper Eocene and Oligocene rocks on the entire Pacific Coast. In that single long section, we were able to identify all the magnetic chrons from about 37 million to 23 million years ago and to calibrate the Pacific Coast molluscan zones against the global magnetic timescale. That section soon became the "Rosetta Stone" against which future work on the Pacific Coast Eocene–Oligocene transition could be correlated. I didn't realize it at the time, but this trip offered a glimpse of my future.

Molluscan Mystery

The marine Cenozoic rocks of the Pacific Coast were an enormously rich database, yet they were still poorly understood and grossly understudied even a century after they had first been discovered. They were full of marine mollusk fossils (figure 6.5) some of which were extraordinarily beautiful, along with fossils of sea urchins, incredible crinoids, and a panoply of sharks, bony fish, and marine mammals. In some cases, such as the legendary Sharktooth Hill bone bed northeast of Bakersfield, California, they even contained remarkable land mammal fossils that had apparently floated out to sea and been buried in deep water

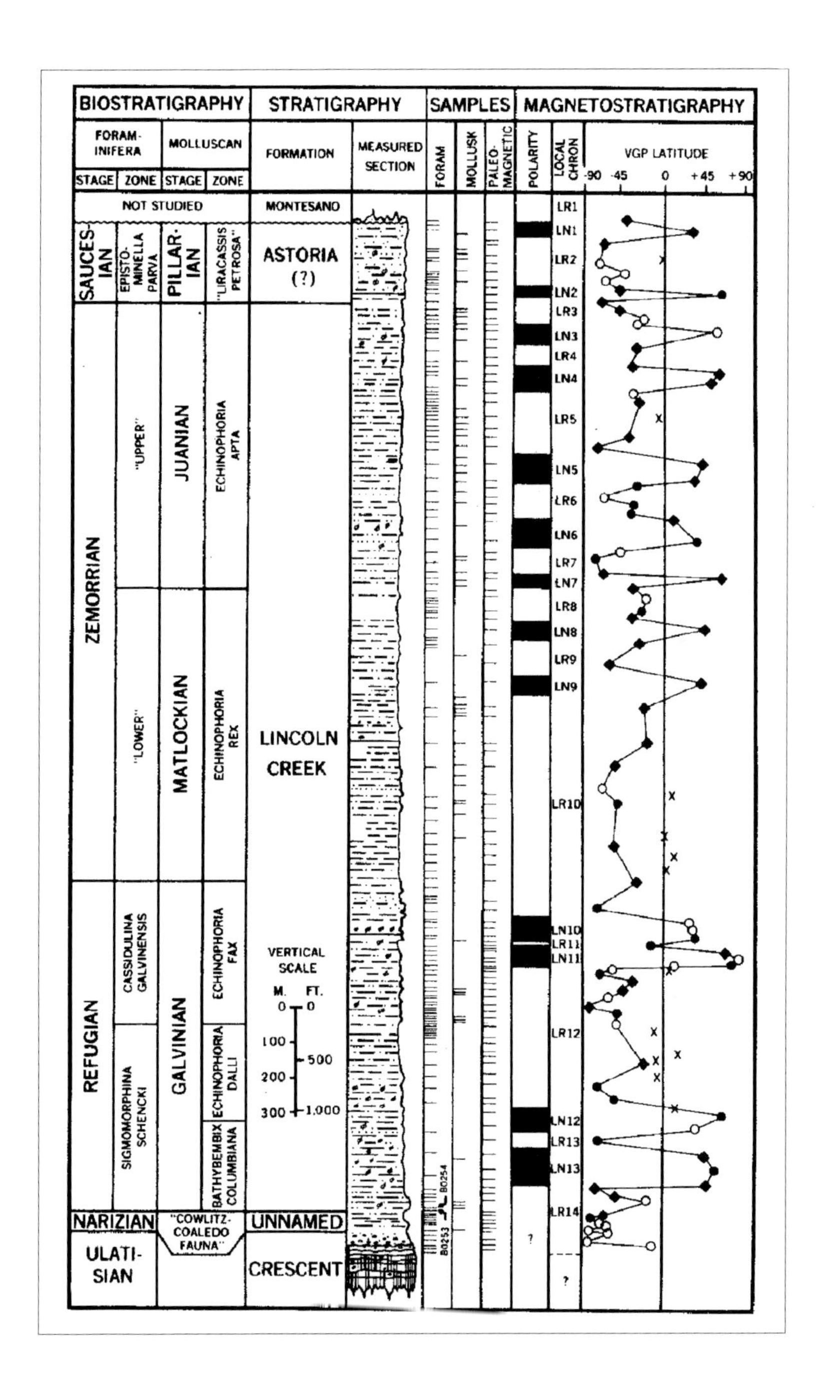

FIGURE 6.4 The published results of the Canyon River paleomagnetic study. (From Prothero and Armentrout 1985)

FIGURE 6.5 In many places, Cenozoic marine beds are rich in fossils: (*A*) thick concretion of the cockle shell *Venericardia bashiplata* from the lower Eocene Bashi Formation, Tombigbee River, Alabama; (*B*) giant scallop fossils from the low-tide exposures of the upper Miocene Empire Formation, Coos Bay, southwestern Oregon. ([*A*] photograph courtesy L. C. Ivany; [*B*] photograph by the author)

along with whales and shark teeth. In the late 1990s, I found myself working at Sharktooth Hill on the private ranch of the legendary collector Bob Ernst. Even though Bob was an amateur, he was also the landowner of the main property containing the bone bed and spent nearly every day collecting out there in the blazing summer heat and cold wintry Sierra winds. He would hire a bulldozer to carve away all the rock overburden that covered the bone bed, then work on a broad bench right on top of the bone-bearing layer, carefully digging out shark teeth, whale vertebrate and ear bones, and the teeth of the huge sea lion *Allodesmus*. Many other marine mammals, including manatees, desmostylians, and early seals, were also found there. Over the years, he and other Sharktooth Hill collectors had located a surprising number of middle Miocene land mammal fossils, including carnivores, horses, camels, deerlike dromomerycines, rhinos, and even mastodonts. My colleagues and I have just published papers on the magnetic stratigraphy of these beds and on the fossil land mammals, but there is still a great deal to do (Prothero, Sanchez, and Denke 2008; Prothero et al. 2008). Sadly, my friend Bob Ernst died suddenly in 2006, right before he was scheduled to guide yet another field trip of my eager students to collect in this legendary locality.

This research builds on almost a century of work by previous paleontologists. Many of them were truly legendary. Some early pioneer paleontologists—such as Thomas Condon, Edward D. Cope, Timothy Conrad, William Dall, and Ralph Arnold—had made limited collections and preliminary maps of the geology (Addicott 1981; Orr and Orr 1999), but the modern phase of paleontology in this region began with John C. Merriam, who founded the first paleontology program on the West Coast at Berkeley in 1912. Although Merriam did some work with Cenozoic invertebrates, Bruce Clark, Merriam's student and later colleague at Berkeley, did most of the research, along with Harold Hannibal and Earl Packard.

Charles Weaver was also originally a Ph.D. student of Clark's at Berkeley. Because Weaver did not drive, he would walk 30 to 50 kilometers (18 to 30 miles) to reach his field area around Mount Diablo, sleeping on the ground whenever he had to stop for the night. As the paleontologist Warren Addicott put it, "[E]ven after retirement, [Weaver] had the reputation of being able to walk a much younger man into the ground" (1981:5). Weaver finished his doctorate and moved to the University of Washington in Seattle in 1907, where he founded their paleontology program and began his research on the Cenozoic deposits of western Washington. There he became a legendary teacher, so painfully shy that early in his career he lectured from the back of the room. He was color blind, so he could not read traffic lights, yet somehow he managed to teach optical mineralogy. Weaver was known for his amazing memory and ability to teach all of his courses without notes. Unable to drive and with extremely low budgets, he would hitch rides on trains or in cars to reach his field area, then do all the rest of his work on foot.

Between 1906 and 1916, Ralph Arnold, Bruce Clark, Harold Hannibal, and Charles Weaver laid the foundation for all future paleontological research in the Pacific Coast Cenozoic. They mapped and collected mollusks from most of the major formations, and they had formulated a preliminary timescale by 1916 (Prothero 2001a). By the late 1930s, however, the research on Cenozoic marine rocks and fossils of the Pacific was slowing down because the great initial exploratory phase was over. It culminated with a report issued by a committee headed by Charles Weaver, working with many coauthors. The Weaver Committee report was published in 1944 in the prestigious *Geological Society of America Bulletin*. It became the standard for Cenozoic West Coast biostratigraphy for decades to come.

Despite all this hard work, the Weaver Committee correlations still had some serious problems. The biggest issue was that the marine mollusks of the Pacific Coast were generally very different from the mollusks and other fossils of the classic areas of the Eocene and Oligocene in Europe, so it was difficult to decide where the standard European time terms fit within the Pacific Coast Cenozoic framework. The Weaver Committee report included two very conflicting ideas of where the Eocene–Oligocene boundary should fall in North America and even one very peculiar term, "Eo-Oligocene," showing how confusing and difficult the dilemma was. The problem was simply insoluble using mollusks, no matter how many more were collected.

Microfossils to the Rescue

During the early part of the twentieth century, another field of paleontology was being developed: micropaleontology. Microfossils were originally overshadowed by the flashier and more popular vertebrates and macroscopic invertebrate fossils. But after World War I, the burgeoning oil industry soon found many uses for microfossils. A micropaleontologist needed just a few scrapings of outcrop or a few chips of well cuttings, and he or she could separate out hundreds of microfossils. From these specimens, he or she could date the formations quite precisely and correlate them from well core to well core. In many cases, the microfossils were also diagnostic of the depth of water in which they had lived, so the micropaleontologist could tell the oil geologists that the basin was of a given depth when these fossils had formed and help determine whether the rock was a likely oil source. Legions of micropaleontologists soon populated all the oil companies, especially during the international oil boom after World War II.

The micropaleontologists were naturally also interested in the rocks of the Pacific Coast Cenozoic because these rocks form some of the largest oilfields in the western United States outside Alaska. Following the pioneering work of Joseph Cushman in Massachusetts, who demonstrated the importance of foraminifera for biostratigraphy in the Atlantic Coast in 1914 and in the Gulf Coast in 1922, many other micropaleontologists took interest in the Pacific Cenozoic. The first West Coast pioneers were G Dallas Hanna of the California Academy of Sciences and Hubert Schenck of Stanford.

The most important figure of all was Robert M. Kleinpell. He got his undergraduate degree in 1926 at Occidental College (where I now teach), then worked in the oil companies and became familiar with micropaleontology. He then returned to academia, getting his doctorate from Stanford under Schenck in 1933. His dissertation, representing his work on thousands of oil company drill cores all over California, was entitled "Miocene Stratigraphy of California." After the AAPG published it in 1938, it became the standard reference for all later Cenozoic micropaleontology in this state. Kleinpell documented the best Miocene stratigraphic sequences and their microfossils in California and set up a standard zonation. He also became a major advocate of the standard Oppelian methods of biostratigraphy. His 1938 book laid down many of the principles first articulated by Schenck and now incorporated into the *North American Code of Stratigraphic Nomenclature*. Nearly all California oil company micropaleontologists and stratigraphers adopted Kleinpell's methods from then on.

After finishing his doctorate, Kleinpell went back to working for the oil companies and was sent to the Philippines to explore for oil in the early 1940s. Like many Americans stranded in the Philippines right after Pearl Harbor and the sudden Japanese invasion, he was captured by the Japanese and spent four years in a Japanese prison camp. He kept up his spirits by teaching his fellow prisoners and somehow managed to avoid getting killed or maimed by the brutal Japanese guards. When he returned to the United States, he took a position at Berkeley in 1946, where he trained several more generations of micropaleontologists for the oil industry and for academia. One of them was Stan Mallory, who revived the program founded by Charles Weaver at the University of Washington decades earlier. Mallory, in turn, was the adviser of John Armentrout, who followed in the Berkeley–University of Washington blueprint in many ways. The current paleontologist at the University of Washington Burke Museum is my good friend Liz Nesbitt, who was also trained at Berkeley.

With this nexus of interwoven careers and student-faculty mentoring at Berkeley and Seattle, the Cenozoic micropaleontology of the West Coast made great progress. Using benthic foraminifera, Kleinpell zoned the Oligocene and Miocene of California. A generation later Stan Mallory created a benthic microfossil zonation for the Paleocene and Eocene rocks of California. The Kleinpell (1938) zonation and other micropaleontological research were integrated into the 1944 Weaver Committee timescale, but there were obvious conflicts, especially about the position of the Eocene–Oligocene boundary. At the crux of the matter was the Refugian Stage, which Schenck and Kleinpell had erected in 1936 based on microfossils from the Santa Ynez Mountains west of Santa Barbara. Most micropaleontologists assigned it to the late Eocene, but other paleontologists working on mollusks thought it was early Oligocene (summarized in Prothero and Thompson 2001). Still others compromised and thought that it straddled the Eocene–Oligocene boundary. Because of such conflicts, there was still a great deal of uncertainty in using the Weaver Committee timescale and even more difficulty in deciding how it correlated with events elsewhere in the world.

Amid this confusion, another field of micropaleontology came to the rescue. The early work in micropaleontology focused on the benthic foraminifera, which live on and in the ocean-bottom sediment. They can be relatively large—typically sand-grain size—but sometimes are as large as a quarter, which is enormous for a single-celled organism. They are also abundant and conspicuous in practically any marine sediment. As bottom-dwelling creatures, however, they often prefer certain water depths, so their stratigraphic occurrence in a given section might be due to changing water depths and not just to the evolution of their lineages.

This problem does not plague the much tinier planktonic foraminifera, which are less than a millimeter in diameter and float by the trillions in the shallow waters of all the world's oceans. They were initially poorly understood and understudied because they tended to be rare in the high-sedimentation-rate sections of the marine basins of the Pacific Coast. During the 1960s and 1970s, however, deep-sea drilling began to retrieve hundreds of cores from the deep ocean bottom, and these cores could be dated only by means of planktonic microfossils. The planktonic foraminifera, in particular, became the global standard for time correlation in the Cenozoic (Berggren 1971), as I mentioned in chapter 2. Clearly, the confusion and conflict about Pacific Coast Cenozoic correlations needed experts on planktonic microfossils, such as the foraminifera, as well as on the coccoliths (chapter 5) to resolve the issue.

During the 1960s, a number of micropaleontologists plunged into this task (summarized in Prothero 2001a). It was a very difficult one because planktonic microfossils are often scarce and poorly preserved due to the rapid sedimentation of these Cenozoic formations that are full of mud and sand. Bit by bit, however, micropaleontologists were able to calibrate one formation after another and roughly correlate them to the global timescale. Soon thereafter, Bill Berggren and colleagues (Berggren 1971; Berggren, Kent, and Flynn 1985) used planktonic microfossils to determine how the European type sections correlated with the global microfossil timescale, and they thus laid a firm foundation for the future.

In many cases, the planktonic microfossils conflicted with the correlations based on benthic foraminifera. They frequently showed that the old benthic foraminiferal zones spanned too much time or were different ages in different places.

This difference was due to the now well-known fact that benthics often track the changes in water depth in their local environment, whereas the planktonics do not. For example, work by Jere Lipps, Dick Poore, Alvin Almgren, and Kristin McDougall showed that many of the benthic "zones" used by earlier workers were different ages in different places and held no time connotation over long distances. By the 1980s, planktonic microfossil specialists had recalibrated much of the original Weaver Committee timescale.

Despite these corrections, there were still major problems. Many of the units had only very scarce planktonic microfossils at just a few levels, so the precise dating of the entire formation was impossible. Only a very few units had volcanic ash layers that were amenable to potassium-argon dating, so almost no numerical dates were known to calibrate these units. But these conditions were just the kind in which magnetic stratigraphy could resolve questions that had never been answered before. It was with this thought in mind that I began to plunge into paleomagnetism on the Pacific Coast in the 1980s.

Magnetobiostratigraphy in the Pacific Coast Cenozoic

When I moved to California and started my research on the marine Cenozoic of the Pacific Coast, much work needed to be done. I began in 1986 by working on marine rocks in the mountains around southern California. Each time I went out, I would find another small local magnetic stratigraphy project that would be ideal for one of my students at Occidental College to work on as a senior comprehensives research project. And so, year after year, two or three Oxy undergraduates would work with me: we'd finish another crucial formation, and they'd get a published project to start their own careers on a good track. By 1991, I was done working the Rockies and Plains, and it was time to tackle projects farther up the California coast. By 1997, I had grant money to travel all the way to Oregon and Washington, where I took a field crew of two students and myself to sample some key regions, such as the Keasey Formation of northwestern Oregon. The 1998 field crew (mentioned in the beginning of this chapter) consisted of four amazing students: juniors Clio Bitboul, Linda Donohoo, and Elana Goer and senior Karina Hankins (see figure 6.1). They worked incredibly well together and managed to cope with extremely difficult conditions without ever whining or complaining or slowing down on the job. Thanks to their hard work, we collected a record number of samples on nearly all the key Cenozoic formations we could sample in Oregon and Washington in just more than a month. I did additional sampling in 1999 and again in 2000, 2001, 2002, and 2005, so that most of the major Cenozoic marine formations of the West Coast have now been analyzed paleomagnetically.

After each summer up in the beautiful, cool, comfortable Pacific Northwest, my students and I would return to southern California. There I'd cut down the little oriented hand samples into cubes of rock on a band saw, and we'd work weekends and holidays (especially around Thanksgiving and between Christmas and New Year) to get the samples analyzed on the amazing Caltech magnetometer. When I first started working in paleomagnetism in 1979 and 1980 in the paleomagnetics lab at Woods Hole and in 1983 at the South Dakota School of Mines, the process was very slow and time consuming. I had to load each little cube of rock by hand, then raise, lower, and turn the long tube that brought the sample into the sensing region deep in the magnetometer, making all the measurements and recording them manually or on a small calculator (see figure 3.1). By the late 1980s, the Caltech lab had a motorized sample holder, so that all the raising, lowering, and turning of the specimen as it was measured could be done with just a few computer commands, although the sample had to be changed manually every minute or so.

Then in the late 1990s, Joe Kirschvink and his students at Caltech invented an amazing new technology: a completely computer-driven automated sample changer (figure 6.6). It used small one-inch-diameter cylindrical cores instead of little cubes of rock, so I took the oriented hand samples from the field wrappings and drilled oriented cores out of them, using a drill press in the rock prep lab, which made them the same size and shape as the conventional cores drilled in the field with a portable drill. The long plastic wand of the manual sample-changing operation had been replaced by a long, nonmagnetic, quartz-glass tube, which was hooked to a vacuum. Above the opening of the magnetometer was a small flat table with a Teflon surface on which there was a long "bicycle chain" belt of little cylindrical "links," each just large enough to hold one core sample apiece. Motors would drive this little "bicycle chain" back and forth from the original position to a position under the glass rod. Then other motors would lower the glass rod down onto the sample, and the vacuum would switch on, picking up the sample. After that, the belt moved to an open "link," so the rod could lower the sample through the hole in the chain and the table. Then the motors would drive it through all the measurements and return it to its original place on the belt. The alternating field coil was just below the opening in the little sample-changing table. We could load up to 80 samples to be measured, demagnetized by progressive alternating field steps, and remeasured each time, and all the data were automatically saved to hard drive. If we had many samples and many demagnetization steps, we could program the system to run automatically overnight and then come back the next morning to find all our data ready to analyze.

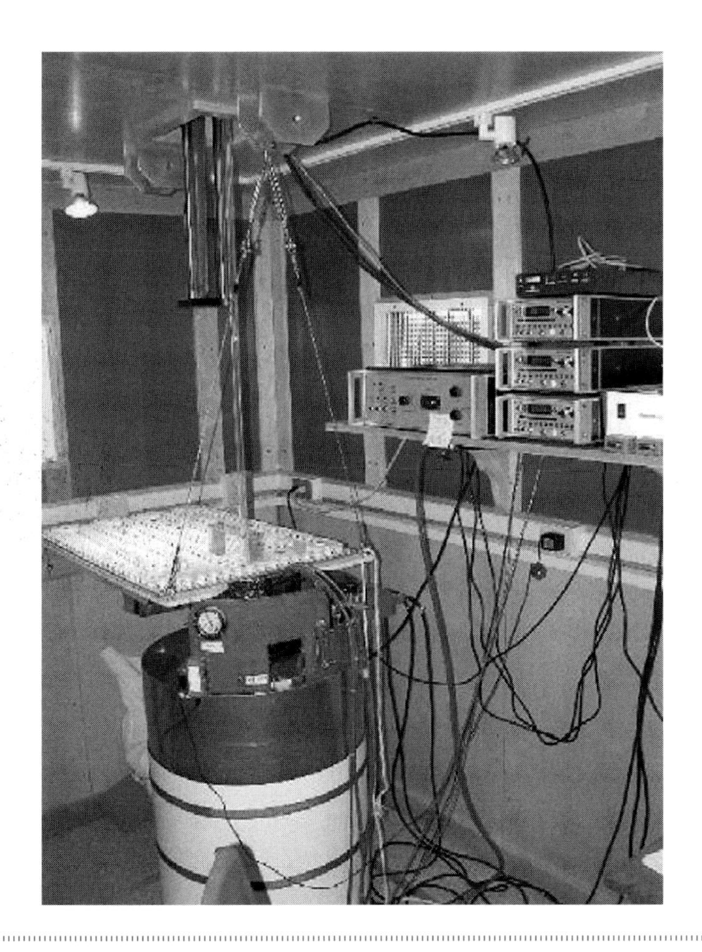

FIGURE 6.6 The 2G Enterprises cryogenic magnetometer with Caltech-style automatic sample changer in the Occidental College paleomagnetics lab. The large cylinder is the dewar of liquid helium, and the cylinder below the table (behind the panel with the dial display) is the alternating field demagnetization coil. The Teflon-topped table has a series of core holders that resemble the links on a large bicycle chain. Computer-driven motors move this chain back and forth so that up to 80 samples can be picked up by the motorized vacuum in the quartz-glass tube above the table, lowered down into the sensing region inside the magnetometer, and turned to measure the specimen in three axes. (Photograph by the author)

This amazing homemade system had a few kinks to iron out in the early phases, but it worked very well overall, and a huge amount of data was collected with minimal labor in a relatively short time. Joe and his students eventually learned from the problems of the old system and invented version 2.0 of the automatic sample changer. This version was mass produced with five complete copies, so that there are now five labs in the world that have this amazing setup: the original Caltech machine, plus three machines in the labs of Joe's former students who are now professors at Yale, MIT, and elsewhere, and the machine at our new state-of-the-art paleomagnetics lab at Occidental College. We are the only small liberal arts undergraduate college in the world with such sophisticated lab equipment. Every year several of our seniors complete and often publish their senior comprehensives research project in this laboratory, and many other visitors from big universities that have no paleomagnetism laboratory come to use our lab.

Over the course of the late 1980s and the 1990s, my student crews and I visited locality after locality in California, Oregon, and Washington, including nearly all the critical exposures of the benthic foraminiferal stages erected by Kleinpell, Schenck, and Mallory. We also sampled most of the classic sections that were the basis for the Weaver Committee's (1944) molluscan stages, including stages erected by Ralph Arnold, Harold Hannibal, Bruce Clark, Charles Weaver, and Harold E. Vokes (by the way, Vokes was also an alumnus of Occidental College, class of 1931). These field areas (figure 6.7) ranged from the Eocene, Miocene, and Pleistocene beds in San Diego County to the long Cenozoic sequences in the Coast Ranges and Transverse Ranges of central California, to the Eocene Vacaville Shale (type section of the Ulatisian) near Sacramento, and to the Cretaceous, Eocene, and Miocene rocks of the Gualala block, west of the San Andreas fault between Point Reyes and Cape Mendocino. We sampled the Eocene–Oligocene Coaledo and Tunnel Point formations, the Miocene Empire Formation in southwestern Oregon near Coos Bay, and a whole series of coastal exposures of Eocene, Oligocene, and Miocene units in northwestern Oregon from Newport and Alsea Bay north almost to Astoria. We also sampled the creekside exposures of the Eocene and Oligocene Cowlitz, Hamlet, Keasey, and Pittsburg Bluff formations in the forests of northwestern Oregon between Portland and Astoria. In Washington, we sampled Eocene, Oligocene, and Miocene units all around the Olympic Peninsula. This sampling included the Lincoln Creek, Humptulips, and Montesano formations on the south side; the upper Oligocene Blakeley Formation on the eastern flank across Puget Sound from Seattle; the Lyre, Quimper, and Marrowstone formations on the Quimper Peninsula of the northeastern Olympic Peninsula; and the Hoko River, Makah, Pysht, and Clallam formations on the northern shore. We've even sampled the age-equivalent Sooke Formation across the border on the southern shores of Vancouver Island, which had long been famous for its nearshore mollusk and marine mammal fossils.

By the end of the 1999 field season, this long campaign of sampling and paleomagnetically analyzing nearly every important Pacific Coast Cenozoic unit from the Mexican border to the Canadian border was reaching a climax. On May 17, 1997, we had held our first symposium presenting this research to the geological community at the Bakersfield meeting of the AAPG and the Society for Sedimentary Geology (SEPM). It was an incredible meeting. Many of my students got up and presented their research for the first time to a real scientific audience, so it

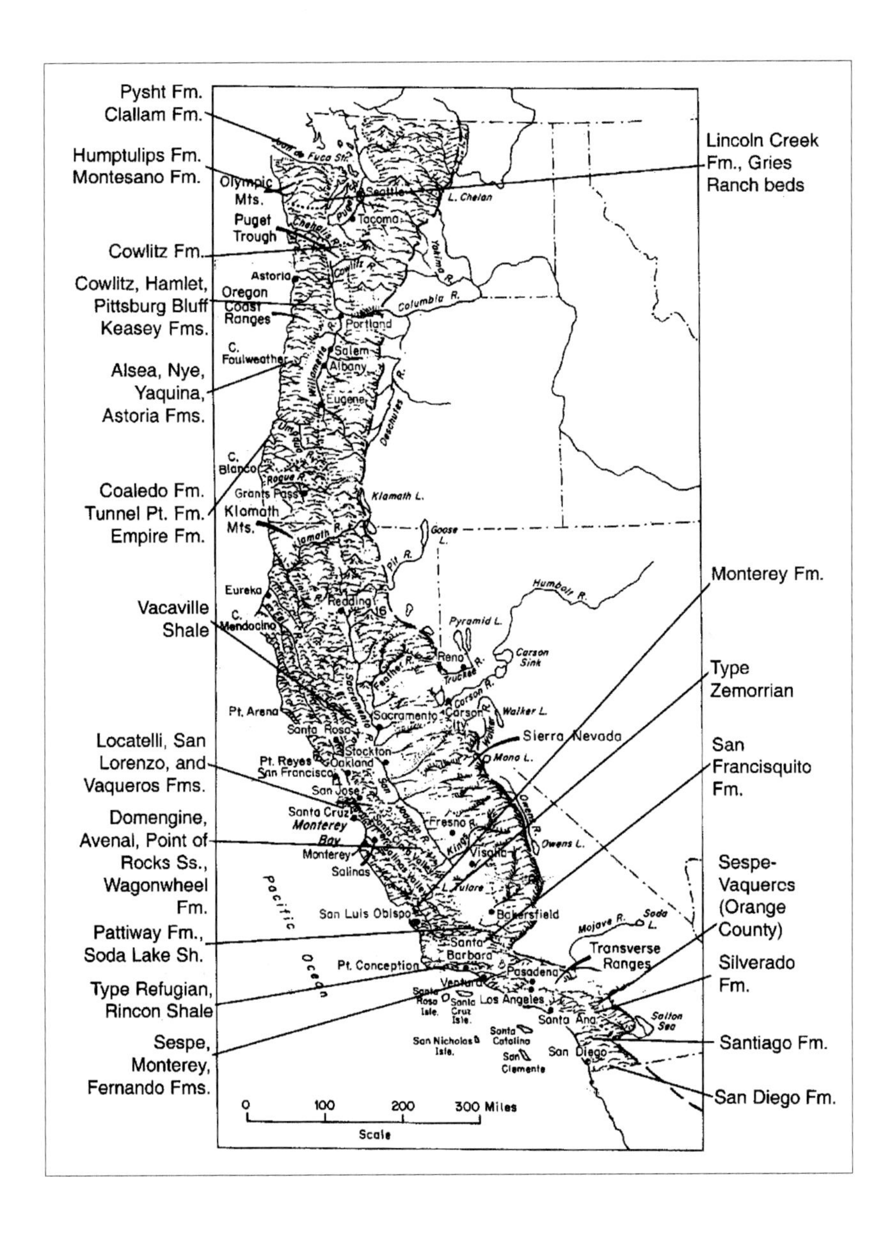

FIGURE 6.7 The location of the major Cenozoic beds that were paleomagnetically sampled, analyzed, and then published in the symposium volume *Magnetic Stratigraphy of the Pacific Coast Cenozoic* (Prothero 2001c).

caught the attention of many of the prominent Pacific Rim geologists who were in attendance. The famous paleomagnetist Jack Hillhouse of the USGS called our work a "tour de force." I was hoping it would also catch the attention of the many oil company geologists who were listening, but by this point the oil companies in southern California were no longer doing much drilling or exploration in this area. They had gone through many years of layoffs, so not many currently employed oil geologists were still doing California geology.

After several more years of intensive research, mostly in Oregon and Washington, we organized a second day-long symposium, this time for the meetings of the Cordilleran section of the GSA and the Pacific Section of the SEPM, to be held at Universal City in North Hollywood on April 10, 2001. Once again my students dominated the program in giving their first professional presentations, and they acquitted themselves magnificently. And once again our research program caught the eyes of the geologists who worked on the Pacific Coast Cenozoic, and we received much praise and helpful feedback.

Some of our results had already been published in the major peer-reviewed geophysical journals such as *Earth and Planetary Sciences Letters*, *Geophysical Journal International*, and *Journal of Geophysical Research*. But what we really needed was a single convenient place to summarize all the older published results and much of our new data. Such a volume would offer not only a complete synthesis of what we had learned since the development of the 1944 Weaver Committee timescale, but also individual papers. I approached the Pacific Section SEPM to see if they would be interested in publishing this volume, and they agreed. Then came the massive task of getting all the results written up in 30 different chapters, getting them peer reviewed and revised, and editing numerous other chapters by paleomagnetists outside my group who wanted to contribute. To save time and money, I produced the entire volume on my own computer, using the QuarkXPress publishing software that I've used to produce several other books camera-ready for the printer. The book finally came out in early 2001. Entitled *Magnetic Stratigraphy of the Pacific Coast Cenozoic* (Prothero 2001c), it has 35 chapters covering almost 400 pages. In this volume is literally everything you ever wanted to know about the dating of the Pacific Coast Cenozoic.

In his review of the book, Bill Berggren wrote:

> It was the saturation approach taken by Don Prothero, his students in the 1990s that finally brought the whole Pacific Coast Cenozoic record under scrutiny with the integrated magnetobiostratigraphic approach that has proved so fruitful elsewhere.... The task Prothero and his colleagues set themselves was formidable: no less than a comprehensive magnetic sampling of all the major Paleocene-Pleistocene outcrops from San Diego to the Straits of Juan de Fuca.... The results of this volume are voluminous and of great importance to placing

the Pacific Coast Cenozoic stratigraphic succession into an ordered sequence, demonstrating the relationship between time and stratigraphy in a given section.... The integrated studies on magnetobiostratigraphy by Prothero and his colleagues have revolutionized our perception of Pacific Coast Cenozoic stratigraphy, and, as the saying goes, "things will never be the same again." (2002:528)

Marine Synthesis

Why would anyone spend 20 years of his life paleomagnetically dating all these rocks from the Pacific Coast Cenozoic? Who would care? It turns out that the now-depleted ranks of Cenozoic stratigraphers in the oil companies were very interested, and they've used our research to find more oil. But there was a much more scientifically interesting reason for the work as well. The deep-sea marine record for the climatic changes of the Eocene and Oligocene is extremely well documented. Forty years of drilling cores all over the world's oceans have produced an incredibly detailed record of microfossils and their changes due to climate, as well as the geochemical changes in the ocean waters in response to climate. Almost all studies of paleoceanography and paleoclimate are built on this foundation, and hundreds of people work in this area of research.

A much smaller number of paleontologists and stratigraphers work on shallow marine rocks with their fossil records of the changes in mollusks, sea urchins, corals, and the rest. Some work in Europe, which was under the ocean during much of the early Cenozoic. Europe is the classic area where the subdivisions of the Cenozoic timescale were first named and established. For many years, intensive work has been done on the Eocene and Oligocene beds of the Gulf Coast of the United States, which runs from coastal Texas through Louisiana, Mississippi, and Alabama, and into Georgia, the Carolinas, and Florida. The Gulf Coast Eocene and Oligocene are world famous for their amazing shell beds (see figure 6.5A) in legendary localities such as Little Stave Creek, Alabama. This place is so dense with beautiful, perfectly preserved shell and shark teeth fossils that you can collect them by the bucket with just a trowel. I got a chance to visit many of these famous localities on a GSA field trip during a meeting in New Orleans in 1995. It was staggering to see the quantity and quality of the shells there! Legendary paleontologists such as Katherine van Winkle Palmer and L. D. Toulmin had documented thousands of species of snails and clams from these beds, and these fossils were currently under study by my friend David Dockery of the Mississippi Office of Geology in Jackson. Abundant corals, echinoderms, and especially benthic foraminifera could also be found in these beds, so it was a very important record of Cenozoic marine evolution.

By comparison, there had been much less recent work on the fossils of the Pacific Coast Eocene and Oligocene (Orr and Orr 1999) because there were fewer paleontologists who worked on them, and, in general, the Pacific specimens were typically enclosed in hard stony matrix and not well preserved, so they could not be easily extracted and studied like the loosely aggregated and perfectly preserved shells of the Gulf Coast. Yet there was still enormous potential to summarize the huge amount of work that had been done on Pacific Coast Eocene and Oligocene fossils if only they could be precisely correlated to see if their evolutionary and faunal changes were in synch with those from the Gulf Coast and with the global oceanographic changes that the marine micropaleontologists had documented.

That's where all the Pacific Coast paleomagnetic dating that my students and I did comes in. With the precise magnetobiostratigraphic correlations to the nearest 100,000 years on the global timescale, West Coast paleontologists could compare their record of faunal and climatic change at high resolution with the data from the Gulf Coast, from Europe and the rest of the world, and especially from the deep-sea cores. It was possible to synthesize all this information for the first time and see what we could come up with.

I'd been in touch with the major players of Pacific Coast Cenozoic stratigraphy and paleontology for a long time as I worked in one research project after another, so they were kept up on the major results as soon as they came out of the magnetometer. But what we really needed was a meeting where we could bring together paleontologists and stratigraphers who worked on the Eocene and Oligocene of the Pacific, Gulf, and Atlantic coasts, along with those who worked in Eurasia and elsewhere in the world. We also needed a contingent of the micropaleontologists who knew the latest thinking about paleoclimatic and faunal changes in the world's oceans. We needed another Penrose Conference.

In chapter 3, I described the experience of my first Penrose Conference in Rapid City, South Dakota, in August 1989. That meeting had a wide spectrum of different types of geological and paleontological expertise. There were many specialists on land mammals and rocks, but only a small minority who studied the marine Eocene–Oligocene record. A decade later we had made so much progress in understanding the marine Eocene–Oligocene that it was time for a more focused Penrose meeting. I got together with my friend Liz Nesbitt of the Burke Museum, an expert on Eocene mollusks of the West Coast, and Linda Ivany of Syracuse University, an expert on the Gulf Coast mollusks and formerly a student of Stephen Jay Gould at Harvard, and we made plans to organize the meeting and propose it for approval by the GSA Penrose Committee.

Once the proposal was approved, we brainstormed when and where to hold the conference. Because most attendees were academics who taught during the school year, the best time for the meeting was in the summer of 1999, but holding it somewhere along the Gulf Coast during its worst period of August heat and humidity just so we could have a field trip to some important localities would be insane. We considered meeting in Ventura or Santa Barbara, California, where many of my classic magnetic sections were sampled, but neither the collecting nor the quality of the mollusk fossils were good, and that area can be blazing hot in August once you get away from the coast. Instead, we agreed that Liz would host the meeting in Washington, where it is typically cool and dry during August, a great relief from the heat waves that cooked the rest of the country. Liz and GSA Penrose coordinator Lois Elms found that Evergreen State College, a lovely little school in Olympia surrounded by deep forests, was available at that time, so we booked the conference room in the student union and lodging in Evergreen's wonderful dorms that look like rustic cabins. We shared these facilities with the University of Washington Huskies football team, which held their summer training camp there. Quite a juxtaposition at the cafeteria each meal: huge, muscular football players in shorts and T-shirts mixed with much smaller, mostly scholarly, bespectacled paleontologists!

Sure enough, the Penrose magic happened again. A wonderful group of 47 scientists (figure 6.8) from the United States, Canada, Cuba, China, Japan, Australia, New Zealand, Germany, Belgium, and Hungary attended. Among them were nine graduate students and five Occidental undergraduate students (Clio Bitboul, Karina Hankins, Linda Donohoo, Ashley Streig, and Elizabeth Sanger) who came with me and presented their research for the benefit of the

FIGURE 6.8 Group shot of the participants in the 1999 Penrose Conference on the marine Eocene–Oligocene transition, held in Olympia, Washington. (Photograph courtesy A. Oleinik)

conference. In the workshop atmosphere, we alternated between formal talks and informal discussion, with occasional breaks for poster presentations. In the evening, there were lively mixers where the scientific discussions continued until late at night. Midway through the meeting, we had a field trip to get a break from dark rooms and all that talking. Liz Nesbitt, John Armentrout, and I organized a nice relaxed excursion to classic localities in the nearby Lincoln Creek and Cowlitz formations, where everyone got their share of collecting nice Eocene and Oligocene fossils and saw the typical Pacific Coast exposures along road cuts and creek bottoms. One night we had a wonderful dinner of salmon roasted Tlingit-style in a great outdoor barbecue pit, where we could watch the sun set on Puget Sound. By the end of the meeting, many important scientific discoveries had been communicated and a few breakthroughs made as well. More important, we established many friendships and scientific collaborations that have lasted for years. Among these connections, my student Linda Donohoo met and fell in love with her future husband John Hurley, a grad student from Ball State University in Muncie, Indiana. Today they live in Albuquerque where John works and Linda is finishing her Ph.D. I knew that Penrose Conferences were great opportunities for people to meet, but never did I expect to become a matchmaker!

The meeting opened with a keynote address by the legendary micropaleontologist and paleoclimatologist Jim Kennett, who updated us on the new discovery that methane hydrates in the seafloor sediment may be a key to understanding Cenozoic climates. These methane hydrates are molecules of methane (natural gas) locked up in cages of ice that are trapped in sediments in the deep ocean where it is cold and under great pressure. Many scientists now think that if ocean-bottom temperatures warm past a critical threshold, these "ice cages" melt, possibly causing a huge amount of the greenhouse-gas methane to be released to the atmosphere.

The Penrose Conference had a great diversity of scientific expertise. Pacific Coast mollusk specialists Dick Squires, Carole Hickman, and Liz Nesbitt summarized their work on Eocene–Oligocene mollusks of the United States, and Anton Oleinik and Louie Marincovich described mollusks from Kamchatka and Alaska. All of them noted a remarkable trend, with typically tropical middle Eocene mollusks vanishing or retreating during the late Eocene cooling, and polar mollusks from Siberia and Alaska appearing in California by the Oligocene. We heard from Dave Scholl about remarkable glacial deposits on the floor of the North Pacific, possibly related to the early Oligocene glaciation. Aussie Paul Gammon told us about remarkable siliceous sponge beds in southern Australia that may have been due to the development of the modern circum-Antarctic current, which brought abundant silica to the shallow waters off southern Australia. Ewan Fordyce from

New Zealand noted that the latest Eocene and early Oligocene represented the first appearance of modern toothed and baleen whales, which quickly radiated in the Oligocene possibly due to the more vigorous water circulation and the bloom of plankton around the Antarctic.

Our second day of meetings focused on the Atlantic and Gulf coasts of the United States and the Caribbean, where a number of workers reported on the changes in mollusks, echinoderms, fish, and benthic foraminifera as these regions cooled dramatically through the middle and late Eocene and into the Oligocene. Linda Ivany presented her fascinating research into the geochemistry of the ear bones of conger eels, which documented the annual changes in temperature as these bones evolved from 40 to 30 million years ago (see pp. 187–189).

After our midmeeting field trip, we spent our third morning with a summary of the global record from the open oceans by Liselotte Diester-Haass and Ellen Thomas, followed by presentations on the record from Europe and Asia. The final morning allowed the impact advocates such as Wiley Poag to present their case, although by this time the overall message from the previous talks was that there was no significant extinction associated with the 36.5-million-year-old Chesapeake and Popigai impacts. We spent our final afternoon in wide-ranging, free-for-all discussions moderated by John Armentrout. Here we really let our hair down and let the spirit and camaraderie of the meeting take us to new levels, where a clear message and a summary of our conclusions emerged. All in all, the five-day conference was wonderful if exhausting, and through it we accomplished more in just a few days than most meetings and scientific papers accomplish in a decade.

The final step was to get all this outstanding new research published. It took four years of soliciting chapters from authors, getting reviews, getting authors to revise their chapters, editing, and getting a publisher before the book was finally in print. Our original publisher made unreasonable editorial demands, so we withdrew the book, which slowed the process down by more than a year. *From Greenhouse to Icehouse: The Marine Eocene–Oligocene Transition* was finally published by Columbia University Press in 2003 (Prothero, Ivany, and Nesbitt 2003). Now a published record of our latest understanding of the marine Eocene–Oligocene is available for anyone who is interested.

Of course, science marches on, and there will always be new discoveries about the Eocene–Oligocene marine record to tantalize and challenge us. But I'm proud of the fact that we established a scientific benchmark for research on that topic as the twentieth century ended and that we laid a firm foundation for all future work that greatly surpassed what was known in 1990. I can just imagine pioneers Clark, Weaver, and Kleinpell smiling, and I hope that our work has done justice to the foundation they themselves laid almost a century ago.

Further Reading

Addicott, W. O. 1981. Brief history of the Cenozoic marine biostratigraphy of the Pacific Northwest. In J. M. Armentrout, ed., *Pacific Northwest Cenozoic Biostratigraphy*, 3–15. Geological Society of America Special Paper no. 184. Boulder, Colo.: Geological Society of America.

Armentrout, J. M., ed. 1981. *Pacific Northwest Cenozoic Biostratigraphy*. Geological Society of America Special Paper no. 184. Boulder, Colo.: Geological Society of America.

Berggren, W. A. 2002. Review of *Magnetic Stratigraphy of the Pacific Coast Cenozoic*, by D. R. Prothero. *Palaios* 17:527–529.

Berry, W. B. N. 1999. Stratigraphic paleontology: From oil patch to academia. In E. M. Moores, D. Sloan, and D. L. Stout, eds., *Classic Cordilleran Concepts: A View from California*, 267–271. Geological Society of America Special Paper no. 338. Boulder, Colo.: Geological Society of America.

Kleinpell, R. M. 1938. *Miocene Stratigraphy of California*. Tulsa, Okla.: American Association of Petroleum Geologists.

Kleinpell, R. M. 1980. *Miocene Stratigraphy of California Revisited*. American Association of Petroleum Geologists Studies in Geology no. 11. Tulsa, Okla.: American Association of Petroleum Geologists.

Mallory, V. S. 1959. *Lower Tertiary Stratigraphy of the California Coast Ranges*. Tulsa, Okla.: American Association of Petroleum Geologists.

Orr, E. L., and W. N. Orr. 1999. *Oregon Fossils*. Dubuque, Iowa: Kendall-Hunt.

Prothero, D. R. 2001a. Chronostratigraphic calibration of the Pacific Coast Cenozoic: A summary. In D. R. Prothero, ed., *Magnetic Stratigraphy of the Pacific Coast Cenozoic*, 377–394. Society for Sedimentary Geology (SEPM), Pacific Section, Special Publication no. 91. Tulsa, Okla.: SEPM.

Prothero, D. R., ed. 2001b. *Magnetic Stratigraphy of the Pacific Coast Cenozoic*. Society for Sedimentary Geology (SEPM), Pacific Section, Special Publication no. 91. Tulsa, Okla.: SEPM.

Prothero, D. R., L. Ivany, and E. Nesbitt. 2000. Penrose Conference report: The marine Eocene–Oligocene transition. *GSA Today* 10, no. 7:10–11.

Prothero, D. R., L. C. Ivany, and E. A. Nesbitt, eds. 2003. *From Greenhouse to Icehouse: The Marine Eocene–Oligocene Transition*. New York: Columbia University Press.

Weaver, C. E. 1916. *Tertiary Faunal Horizons of Western Washington*. University of Washington Publications in Geology no. 1. Seattle: University of Washington.

Weaver, C. E. 1942. *Paleontology of the Marine Tertiary Formations of Oregon and Washington*. University of Washington Publications in Geology no. 5. Seattle: University of Washington.

Weaver, C. E., R. S. Beck, M. N. Bramlette, S. A. Carlson, L. C. Forrest, F. R. Kelley, R. M. Kleinpell, W. C. Putnam, N. L. Taliaferro, R. R. Thorup, W. A. Ver Wiebe, and E. A. Watson. 1944. Correlation of the marine Cenozoic formations of western North America. *Geological Society of America Bulletin* 55:569–598.

Giant redwood stumps at Florissant Fossil Beds National Monument, Colorado. The students on the bridge are from my 1991 field crew: Erin Wilson (*left*) and Walter Lohr (*right*). (Photograph by the author)

7 | Rocky Mountain Jungles and Eels' Ears

When the mountains are overthrown and the seas uplifted, the universe at Florissant flings itself against a gnat and preserves it.

—DR. ARTHUR C. PEALE, GEOLOGIST ON THE HAYDEN EXPEDITION, 1873

Jungles in the Colorado Rockies

June 1, 2003. My crew and I are hiking in the high meadows of the Rocky Mountains in Florissant National Monument, Colorado. We're at an elevation more than 2,600 meters (8,600 feet), and we're huffing and puffing as we hike along in such a thin atmosphere, even though we're not climbing any steep slopes. We're surrounded by typical high-altitude mountain vegetation of the Colorado Rockies: dense groves of ponderosa pine, aspen, fir, and spruce, plus shrubs and grasses that grow only in the summer. In the winters, this area is under many feet of snow, and during the summers the snow doesn't melt completely away from the high peaks. (As I write this in December 2007, the temperatures are in the –20°C range, and the peaks are being hammered by blizzards.) Even in the warmest days of summer, the air is cool and dry. This June we have to wear relatively warm clothes and raincoats because the mountain thunderstorms constantly threaten to soak us and end our expedition for the day. Indeed, just after we finish collecting our last paleomagnetic samples, the storms cut loose and turn our localities into mud pits. We beat a retreat to the rustic town of Florissant, where we settle into a cozy warm, old clapboard-sided restaurant that has changed little since the early days of the first settlers. Here, next to a blazing fire, we dry off, eat, and get warmed up before heading off to our next project. (Ironically, the town of Florissant, Colorado, did not get its name from the fossil floras, but instead was named so because its founder was from Florissant, Missouri).

Herb Meyer, the park paleontologist at Florissant, has been leading us from one exposure to another so we can collect paleomagnetic samples and help date these rocks better than ever before. Yet even as we leave the visitor's center, we see a striking mismatch between the present-day vegetation and climate, on the one hand, and the climate of the geologic past, on the other. Just outside the visitor's center are several huge fossil stumps of giant sequoia trees more than 4.5 meters (14 feet) wide, identical to those found in the cool-temperate redwood forests of coastal California today (see the photograph that opens this chapter). In the visitor's center, display cabinets show hundreds of specimens of fossil leaves, most of which represent plants that occur today in temperate climates: beeches, poplars, willows, soapberry, sumac, firs, pines, spruce, cypress, elms, hackberries, cottonwoods, mountain mahogany, hawthorn, apples, plums, as well as abundant cattails, horsetails, and mosses that grew in this shallow mountain lake basin 34.07 million years ago. Some of these fossil plants belong to groups that are no longer present in Colorado or even in North America. *Sequoia* lives today only on the California coast and mountains. *Koelreuteria* (raintree) and *Ailanthus* (tree-of-heaven) are now found in eastern Asia. More than 140 plant species, known mainly from exquisitely preserved leaf fossils, can be found at Florissant, along with hundreds of species of fossil insects and other delicately preserved fossils that are compressed in these papery thin lake shales. Some of the fossil insects, such as infamous tsetse flies, are today known only from warm climates, such as tropical Africa. Altogether, more than 1,700 species have been reported from Florissant, making it one of the richest localities in the world.

As we hike along, however, we can see that none of these ancient species (except for a few of the conifers and the mountain mahogany) grows here now. The modern climate, especially with its long season of snow and freezing temperatures, is far too harsh for apples, plums, beeches, poplars, willows, soapberry, and sumac. Paleobotanists have known for decades about this mismatch and puzzled over the implications. In his book-length description of the Florissant fossil plants, the late great paleobotanist Harry MacGinitie (1953) discussed the modern analogues for the Florissant plants and speculated about why it was so warm so high in the mountains during the late Eocene. Most paleobotanists make the obvious connection based on the redwood stumps: this Florissant flora grew in a climate that was wetter and warmer, much like coastal northern California today. They estimate that the mean annual temperature then was about 12.5°C (55°F), much like coastal Oregon or northern California, with some plants that cannot tolerate freezing temperatures giving us a lower limit for the cold months in this region.

My crew and I are in Florissant because we're tackling a new phase of understanding the Eocene–Oligocene transition and the change from a greenhouse to an icehouse Earth. Some of the strongest evidence for ancient temperatures and climate on land comes from the land plants. As most people know, plants are extremely sensitive to the climatic conditions in which they grow, so they are powerful tools for paleoclimatology—much more so than are fossil mammals. In addition, most of the fossil plants known from the Eocene still have modern representatives living in well-documented climatic conditions today, so it is not a great leap of faith to estimate ancient climates based on the presence of modern plants with known tolerances. In addition, paleobotanists have yet another tool besides the simple presence or absence of certain key plant species. It turns out that the edges of leaves are also a good predictor of temperature and climate as well. Tropical leaves have smooth (or "entire") margins and often are also much larger, thicker, and nondeciduous, and in rain forests they have drip tips to help drain off water (figure 7.1*A*). Leaves from cooler climates are smaller, thinner, and often deciduous (except for evergreens), and they typically have jagged edges. Paleobotanists have long puzzled over why the presence of jagged edges is such a good predictor of climate, but the relationship is well documented and very convincing. Dana Royer and Peter Wilf (2006) showed by experiments that tooth-margined leaves tend to have greater surface area along their edges for photosynthesis and transpiration, which helps them grow faster at the beginning of their short growing seasons. Whether this explanation is the entire story or not, paleobotanists for years have been plotting the ratio of entire-margin to jagged-edged leaves against known mean annual temperature of the modern flora and then extrapolating this linear relationship to the geologic past (figure 7.1*B*).

One of the chief proponents of this method was the legendary paleobotanist Jack Wolfe, who passed away in 2005 after a long career with the USGS. He was a brilliant paleobotanist, even if a bit cantankerous. He described and analyzed more floras than anyone of his generation and even named a series of North American land plant stages to complement the North American land mammal ages. In 2000, he received the highest possible accolade, the Paleontological Society Medal for lifetime achievement. I knew Jack from professional meetings and from editing his paper in a volume I put together, so he was courteous and friendly to me, but never easy to get to know. As early as 1967, Wolfe and David Hopkins had published a paper where they had plotted the difference between late Eocene and early Oligocene floras as an abrupt drop from about 60 percent to about 28 percent entire margins. According to the leaf-margin curves, this drop correlated to an abrupt drop in mean annual temperature from about 20°C (68°F) to about 10°C (50°F), which Wolfe thought was very rapid; in their words, "the cooling took place within an interval no longer than two million years and possibly as short as one million" (Wolfe and Hopkins 1967:73).

In a 1978 paper on the topic, Wolfe showed this big temperature drop as the most abrupt and extreme change in temperature in the entire Cenozoic and named it the Terminal Eocene Event (TEE) (figure 7.2). During the heyday of the impact hypothesis in the 1980s, the TEE received a great deal of publicity and

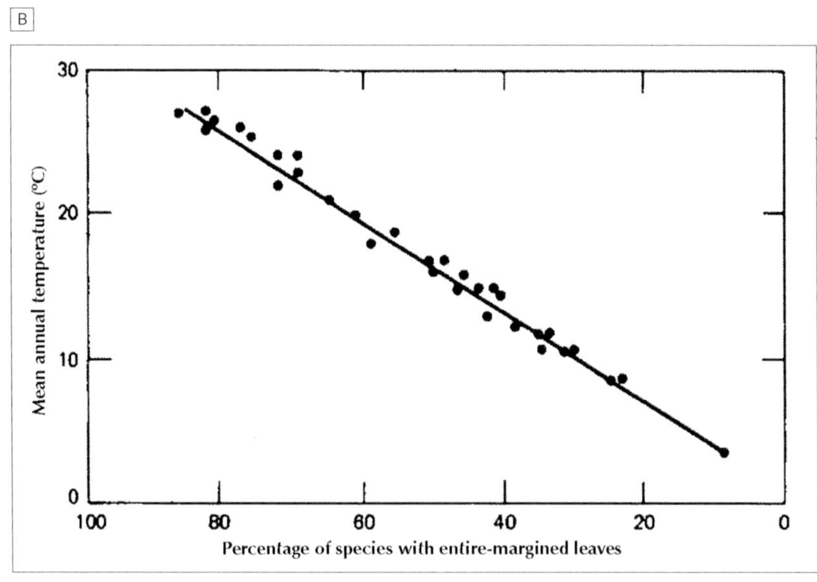

FIGURE 7.1 (*A*) Plants that live in colder seasonal climates have smaller, thinner, deciduous leaves with jagged margins, while plants that live in warm, tropical climates tend to have larger, nondeciduous leaves with smooth ("entire") margins and drip tips; (*B*) the strong linear relationship between the percentage of species with entire-margined leaves and mean annual temperature. (Modified from Wolfe 1978)

was treated as catastrophic, like the KT event (chapter 5). There was even a book entitled *Terminal Eocene Events* (Pomerol and Premoli-Silva 1986). However, during the 1980s, those of us who worked on the Eocene and Oligocene were able to establish that the climatic change was not abrupt or catastrophic and that the biggest change occurred in the earliest Oligocene, *not* at the Eocene–Oligocene boundary. Ironically, Wolfe (1971) had already coined a more fitting term, the *Oligocene deterioration*, and this label is still appropriate for talking about the events of the early Oligocene, when the first glacial ice caps appear in Antarctica and the modern icehouse world began.

So this is the framework that paleobotany had established for the rest of us to follow, but there were serious questions about the data that Wolfe had interpreted. Most of Wolfe's dating from the 1960s, 1970s, and 1980s was very crude, so there was no way to tell how rapidly climatic change had occurred and whether it matched with the global climatic events in the Eocene. At the 1989 Penrose Conference in South Dakota, Wolfe repeated his 1981 argument that the TEE was only about 34 million years old and had arrived in Rapid City to find that our new argon-argon dates agreed with him and ended a decade of conflict (chapter 3).

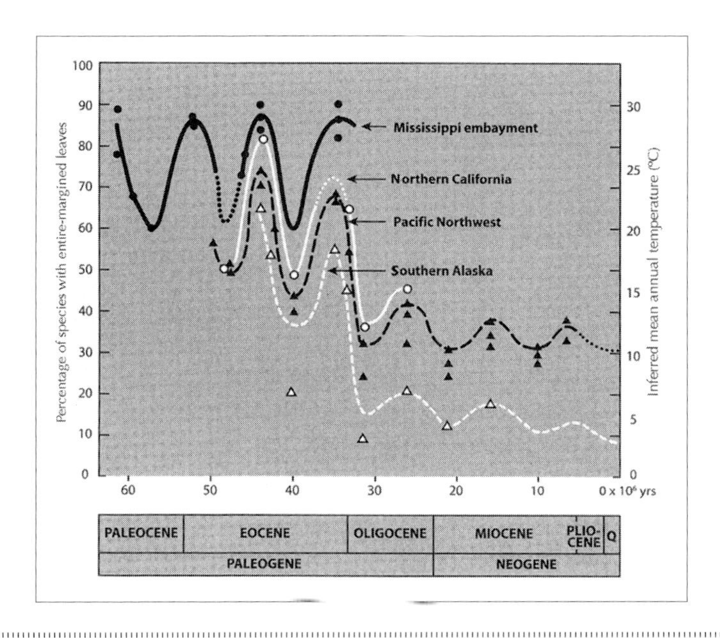

FIGURE 7.2 Jack Wolfe's (1978) climatic curve for the Cenozoic, using the percentage of species with entire-margined leaves as a proxy for temperature. In this interpretation, the drop in temperature during the Eocene–Oligocene transition was large and rapid, based on Wolfe's temperature and age estimates of the floras. This big temperature drop was apparent in the floras of the Mississippi Embayment, northern California, the Pacific Northwest, and southern Alaska. (Courtesy Sigma Xi)

Nevertheless, there were many questions about the dates of most of his floras. In addition, much of his work on the floras had not yet been replicated by anyone else, so there was some question as to whether the drop in temperature was as extreme as he indicated.

By the beginning of twenty-first century, it was time to resolve this problem and get better dates for as many crucial floras as possible. Thus, as I was winding down my late 1990s work on the Pacific Coast marine Cenozoic, I found myself visiting new localities that were famous for fossil plants rather than for mammals or mollusks.

Land of Goshen

So if Florissant shows that the high Rockies were like coastal Oregon in the late Eocene, what about coastal Oregon? Once again we are blessed with some incredible plant fossil localities that give us a startling picture of the climate of the past. In August 2001, my family, field crew, and I were in Eugene, Oregon, visiting my friend and colleague Greg Retallack, a professor at the University of Oregon. Greg is an amazing person; a transplanted Aussie who has lived in the United States for the past 30 years, he has never lost his Down Under accent. His craggy, handsome face, cleft chin, and short wavy hair remind one of the cartoon character Dudley Do-Right of the Mounties in the old *Rocky and Bullwinkle* cartoons, yet behind those looks is an incredible intelligence and quick wit. Greg started out as one of the few people who could dig a trench through an outcrop and read the ancient soil horizons that once grew and developed there. From that training and his background in paleobotany, he has become one of the foremost experts in reading ancient soils and leaves and in deciphering what kind of vegetation and climate they represent. He's not only a theoretical scientist, but also a tireless worker, hiking enormous distances to find fossils or to dig trenches in the Badlands and interpret their soils. And he's not afraid to stick his neck out as well, publishing strikingly original ideas about a wide range of paleontological topics. At the professional meetings, he's all charm and wit, and amazing to talk to (especially with his accent). Most people would never guess from his humble, self-effacing, and friendly attitude and lack of pompousness and pretension that he's one of the brightest minds in all of geology and paleontology.

Greg has taken advantage of his location to spend time roaming the Eugene area and clearing up the geology of this legendary region that has yielded so many important Eocene and Oligocene plant and marine invertebrate fossil localities. Prior to his work, the geology was very confused because many of the fossil localities are covered by the typical dense vegetation of coastal Oregon. Greg, my crew, and I searched the road cuts and along the banks of the Willamette River at low water to get good exposures for fossil collecting or paleomagnetic sampling. Some of our exposures were in awkward places such as the riverbanks below the north end of Autzen Bridge, where thousands of rabid Oregon Duck football fans cross the Willamette on Saturdays in the fall to see their team play at Autzen Stadium (figure 7.3). Others were small embankments above the local Pepsi plant parking lot or road cuts out in the middle of nowhere that only Greg recognized. Some were right in the freeway median, so we had to hike into the tunnel under the overpass (keeping an eye out for huge semi trucks blowing by at 80 miles per hour) to find outcrops beneath all the concrete. Still other classic localities that were important decades ago were no longer accessible because they were covered up by the ever-expanding development of houses and businesses. Others had grown over with dense vegetation when people stopped clearing the road cuts. Many of them were covered with wild blackberries, which have nasty spines and thorns that we had to wade through. There was one benefit of the blackberries, though—they provided good eating once we survived the ordeal, and Greg made us blackberry cobbler each night we stayed with him. Now,

FIGURE 7.3 Collecting samples from late-summer low-water exposures of the Eugene Formation on the Willamette River, just east of Autzen Bridge and the University of Oregon campus (*in background*). My student Elizabeth Draus is working in the foreground. (Photograph by the author)

thanks to Greg's hard work (plus, the paleomagnetic work that my students and I did, and radiometric dating by Paul Kester, Bob Duncan, and Cliff Ambers), the stratigraphic framework of the Eugene area is securely nailed down (Retallack et al. 2004). Each of these Eugene-area floras has been precisely dated to less than 100,000 years by combining the magnetic stratigraphy with the numerous well-dated ashes that run through the region.

One of the most striking of these fossil floras is known from road cuts just off Interstate 5 near the tiny crossroads town of Goshen, Oregon. According to Wolfe (1978), it represents one of the last of the warm tropical Cenozoic floras in North America, dated at 33.4 million years old, so it is probably earliest Oligocene in age. Even today, you can find small excavation pits southwest of exit 186 of Interstate 5 and pull out amazing leaf fossils in just a few minutes. And you don't need to be a trained botanist to notice that they are leaves (figure 7.4) of tropical plants: some are very large, and many look like the kinds of plants you might find in the jungles of Central America. They include such exclusively tropical plants as paw-paws and cold-intolerant trees such as the magnolias, as well as oaks, figs, holly, soapberries, laurels, myrtles, ebonies, roses, legumes, and herbs such as borages and heliotropes. When Ralph Chaney and Ethel Sanborn described this flora in 1935, they realized that it came from a subtropical-temperate rain forest. Modern estimates place the mean annual temperature in the range of 20 to 22°C (68 to 72°F), with an annual range of temperature of only 7°C (13°F), much warmer than cold, rainy coastal Oregon today. Thanks to the annual Rose Festival, we know that roses grow well today in Portland, but not paw-paw trees or magnolias! And just to the south of this Goshen flora locality, near Creswell, is the classic Comstock flora, which represents conditions that were even warmer and more tropical than those at Goshen and are dated at 39.7 million years old. Greg found a back road that took us to the abandoned railroad cuts, where this flora was originally collected. Abundant leaf fossils can be found there even now, showing the distinctively tropical character of the middle and late Eocene in the region.

From the coastal jungles of the middle and late Eocene of Oregon and Alaska to the high mountain meadows of Colorado, the fossil floras show again and again that conditions in North America during the middle and late Eocene were much warmer and wetter than they are now. But what happened to these tropical realms? Once again, the answer can be determined by looking at the fossil plants. Just north of the Goshen locality along Interstate 5 (and just south of the Eugene city limits) is another steep road cut that yields the Willamette and Rujada floras (see figures 7.4*C* and 7.4*D*). And on the east banks of the McKenzie River at Spores Point and Coburg Road, just north of Springfield, Oregon, we sampled a steep cliff that yields the Coburg flora, also early Oligocene in age. The difference is remarkable. The Rujada flora, dated at 31.3 million years old, represents a much cooler,

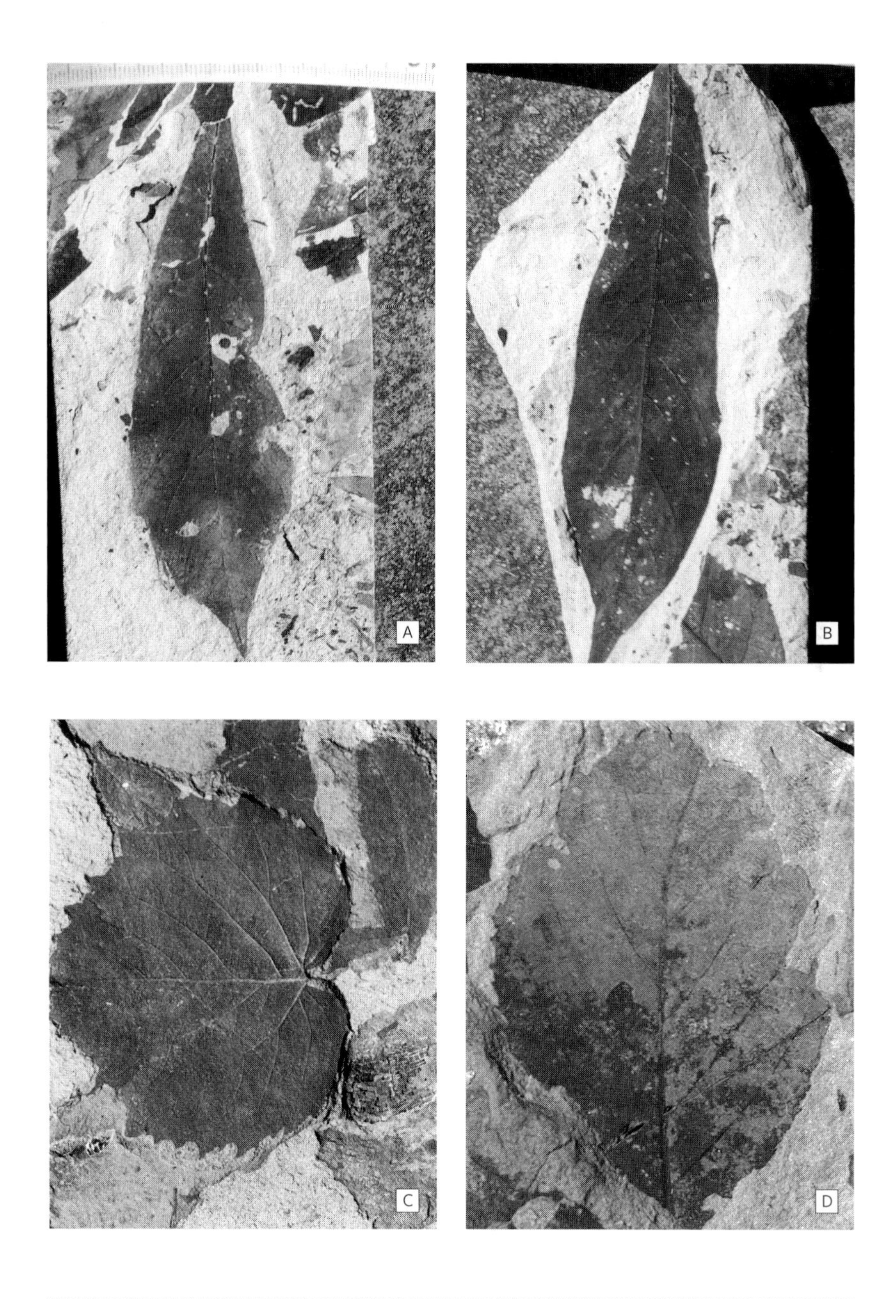

FIGURE 7.4 Typical fossil leaves of the Eugene Formation, showing the changes from the late Eocene to the early Oligocene: (*A*) *Meliosma goshenensis* from the late Eocene Goshen flora, whose living relatives are found in tropical and temperate Asia and the Americas; (*B*) *Laurophyllum merrilli* from the Goshen flora, a laurel best known from the warm early Eocene; (*C*) the plane tree *Platanus condoni* from the cooler early Oligocene Willamette flora; (*D*) the alder *Alnus heterodonta* from the even cooler early Oligocene Rujada flora. (Photographs courtesy G. Retallack)

drier climate, with a mean annual temperature of 13.0°C (55.4°F), a cold-month mean temperature of 2.4°C (36.3°F), a warm-month mean temperature of 23.6°C (74.5°F), and a mean annual range of temperature of 21.2°C (70.2°F) (Retallack et al. 2004). Most of the Rujada plants are typical of cool temperate forests, with abundant leaves of oaks, laurels, alders (see figure 7.4*D*), dawn redwood, ash, hawthorns, sycamores, walnuts, hackberries, elms, lindens, beeches, bracken ferns, and horsetails. Stratigraphically above the Rujada flora are the early Oligocene Coburg and Willamette floras, the former dated at 30.9 million years old and the latter at 30.1 million years old. The Willamette flora grew in climates with a mean annual temperature of 13.2°C (55.8°F), a warm-month mean of 20.8°C (69.4°F), a cold-month mean of 6.2°C (43.2°F), and a mean annual temperature rang of 14.6°C (26.2°F). The less well-sampled Coburg flora is very similar.

In his 1971 and 1978 papers, Wolfe could only speculate how long the Eocene–Oligocene floral transition took place and how extreme it was, but he favored the idea that the change was very large and very rapid (10°C [18°F] cooling in about a million years). Now we can reexamine some of the same floras that Wolfe used, with better dating, and test his hypotheses. Goshen has a mean annual temperature of 19.7°C (67.5°F), slightly cooler than Wolfe's estimate, and Rujada has a mean annual temperature of 13.0°C (55.4°F), so the total cooling is only about 6.7°C (12.1°F), *not* the 10°C that Wolfe and Hopkins (1967) and Wolfe (1978) indicated. Moreover, with the precise dating now available, the age difference between the Goshen and Rujada floras is 2.1 million years, twice as long as Wolfe suggested. Moreover, Wolfe's plots treated the Willamette floras as the same temperature and age as the Rujada floras, but Retallack and his colleagues (2004) showed that Willamette is about 1.2 million years younger than Rujada, with slightly cooler cold-month mean and warm-month mean temperatures. Thus, our newest dating and revised paleotemperature estimates for the Eugene area floras give a more gradual cooling in several steps from Goshen to Rujada to Willamette over about 3.3 million years (figure 7.5). This picture is in striking contrast to Wolfe's interpretation of a catastrophic 10°C cooling between the Goshen and the Rujada/Willamette floras in less than 2 million years (Wolfe and Hopkins 1971; Wolfe 1978).

Back to the Rockies

Let's return to the high Rockies in Colorado and see what happened to the Florissant flora. Once again, we have excellent floras from the Oligocene that show how much things had changed since Florissant time. The best known of

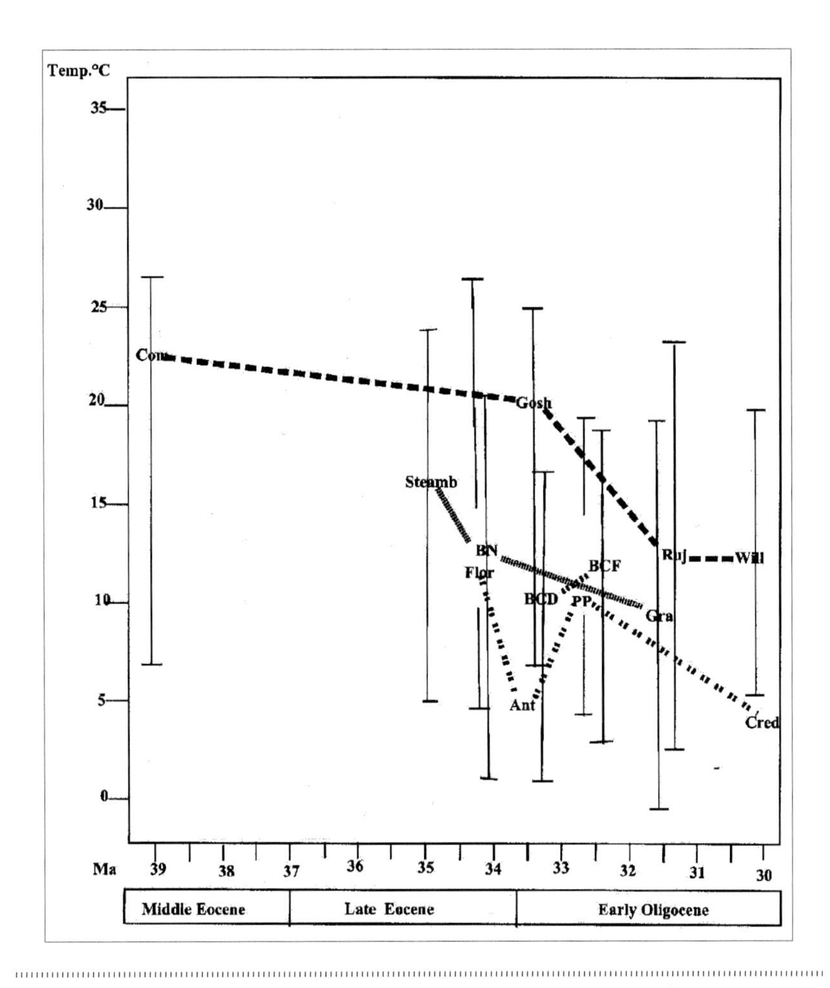

FIGURE 7.5 The change in temperature versus age as represented by the floras discussed in this chapter. The abbreviations are plotted at the mean annual temperature of the flora. The temperature range estimates are shown by error bars above and below each mean temperature. The coastal Oregon floras are connected by the long, bold, dashed lines; the northeastern California floras, by the solid line; the Colorado floras, by the long dashed line; and the central Oregon floras, by the short dashed line. Com = Comstock flora; Gosh = Goshen flora; Ruj = Rujada flora; Will = Willamette flora; Steamb = Steamboat flora; BN = Badger's Nose flora; Gra = Granger flora; BCD = Bridge Creek–Dugout flora; BCF = Bridge Creek–Fossil flora; Flor = Florissant flora; Ant = Antero flora; PP = Pitch-Pinnacle flora (From Prothero 2007b)

them is the Creede flora, from ancient lakebeds that filled the moat of an ancient collapsed volcanic caldera just south of the old mining town of Creede, Colorado. Located in the glacial valleys between the high volcanic peaks of the San Juan Mountains in southwestern Colorado, Creede is an amazing place to visit. The old mining town has been nicely restored with trendy modern restaurants, curio shops, and rock shops, summer-stock theater, and many other activities for tourists. Behind the town, you can still drive a steep and narrow loop road through some of the incredible mine excavations that were in their heyday more than a century ago. But just south of the town is a series of road cuts that tourists ignore as they zoom past. My family and I have explored these cuts often when we spend summers doing research in the Rockies, and every time we visit, we find dozens of beautifully preserved leaves and insects. In the summer of 2005, I collected a paleomagnetic section through these outcrops, and the results came out beautifully when I analyzed them in our lab later that summer. The Creede volcanics have been argon-argon dated between 26.26 and 26.92 million years old, consistent with the magnetic signature I obtained. According to Jack Wolfe (Wolfe and Schorn 1989; Wolfe 1992), Creede leaf fossils indicate much cooler and drier conditions at that time, with a mean annual temperature of only 4°C (39°F) and subfreezing winter temperatures. The flora is dominated by pines, spruce, and junipers. There are also patches of chaparral dominated by broadleaf shrubs such as mountain mahogany, hawthorn, barberry, and Oregon grape.

But Creede is late Oligocene in age (almost 27 million years old), more than 7 million years younger than the Florissant flora. What about the interval in between? A few additional Colorado floras help bridge the gap. In July 2006, I sampled one of these floras, described by Kathy Gregory from the old mine borrow pits of the Pitch-Pinnacle mine south of Monarch Pass and east of Sargents, Colorado. At higher than 3,000 meters (10,000 feet) in elevation, this area is true high-mountain country. Kathy Gregory and Bill McIntosh (1996) dated the floras between 29.0 and 32.9 million years ago, so it is earliest Oligocene, only slightly younger than the Florissant flora. My paleomagnetic analysis narrowed down the age to 30.5 to 31.0 million years old, a much smaller range than originally suggested (Prothero 2007b). The fossil leaves suggest a mean annual temperature of 12.7°C (54.9°F), compared to the range of 12.8 to 13.9°C (55.0 to 57.0°F) for Florissant, but only 4.2°C (39.6°F) for Creede. The cold-month mean temperature is estimated at 4.5°C (40.1°F), warmer than Florissant; the warm-month mean temperature is 20.4°C (68.7°F), cooler than Florissant's 30.0 to 33.3°C (86.0 to 91.9°F); and the mean annual temperature range is estimated at 18.5°C (65.3°F), compared to 27 to 29°C (80.6 to 84.2°F) for Florissant. This warm-climate flora is only slightly cooler than that found in Florissant, so there was only a slight change in temperature between 34.07 and 32.9 million years ago. Thus, the cooling trend between Florissant (34.07 million years old), Pitch-Pinnacle (30.5 to 31.0 million years old), and Creede (27 million years old) was a gradual one, not an abrupt drop at the Eocene–Oligocene boundary or in the early Oligocene (see figure 7.3). This conclusion is consistent with what Greg Retallack, our colleagues, and I concluded in our 2004 paper about the Eugene area floras (Retallack et al. 2004) and in contrast to Wolfe's interpretation.

But there are puzzles as well. In 1992, Wolfe briefly mentioned fossil floras from the Antero Formation in the South Park Basin west of the Rocky Mountain Front Range (and just west of Florissant). The Antero flora was pretty poorly known, with only about six species, including subalpine pine trees, oaks, blueberry, and mountain mahogany. Wolfe (1992) compared it to the Creede flora and suggested that it indicated a very cool temperate flora much like those plants found in Creede. But the Antero Formation yields volcanic ash dates of 33.77 to 33.89 million years old, barely younger than Florissant! If both the age determinations and the floral analysis are correct, then there was a rapid and abrupt cooling event between 33.89 and 34.07 million years ago, completely different from the interpretation of the Oregon floras that Greg Retallack and I published in 2004. Clearly, more work was needed (as the common science cliché goes).

So we set about trying to solve one puzzle: How good was the age estimate of the Antero flora? In July 2006, Herb Meyer and a group of colleagues met me in Antero Junction, from which we caravanned to a number of road cuts where the Antero plant fossils were still coming out of the rock (mostly mountain mahogany leaf fossils in great abundance). At our final locality, we were excited to find mammal fossils as well, including the lower molar of a brontothere, conclusive proof that the beds were Chadronian or late Eocene in age. Just as our excitement was reaching fever pitch and everyone was scrambling to find more fossils, the typical late-afternoon Colorado thunderstorms rolled in, so we had to hightail it out of there before the lightning struck us or the roads became muddy and impassable. The magnetic results from all the Antero localities were beautiful (Prothero 2007b) and, along with the fossil brontothere tooth, confirmed the previous age determinations.

So if the age dating is good and the Antero floras are indeed just slightly younger than Florissant, then what about Wolfe's (1992) suggestion that they were as cool temperate as Creede? Unfortunately, Wolfe's Antero plant fossil collection was very small and incomplete (six species and about 16 specimens), so it does not provide as much evidence as we would like. When Wolfe died in 2005, all of the huge plant fossil collections that he had built from many localities over a long career with the USGS were not kept by the USGS, but ended up in the Smithsonian. Several paleobotanists have been to the Smithsonian and looked through these specimens, but cannot locate his small Antero collection, so we can't tell what fossils were the basis of his conclusions (beyond the brief paragraph in his paper in the volume that Bill Berggren and I edited in 1992). Fortunately, there are other collections, such as the one briefly mentioned by Christopher Durden (1966) and deposited at Yale Peabody Museum. Durden (1966) provided a slightly longer list of plant fossils, including pines, oaks, elms, mountain mahogany, sequoias, Douglas fir, balloon vine, and willows—a flora not too different from Florissant. In fact, one of the species was the birch *Betula florissanti*, confirming the

Florissant affinities to the Antero flora. University of Colorado graduate student Melissa Barton is currently redoing the analysis of the Antero flora and collecting in the new localities. I suspect that once a more detailed study is done, the floral anomaly will vanish, and it will indeed be shown that Antero was much like Florissant in total floral affinity as well as in paleotemperature.

Much had clearly changed between the warm conditions of the late Eocene (Florissant, Goshen, Comstock) and the cool temperate conditions of the Oligocene (Rujada, Willamette, and Coburg in Oregon; Antero, Pitch, Pinnacle, and Creede in Colorado). And there is still much more to be studied. Up in the remote northeastern part of California, near the tiny town of Cedarville and just west of the Nevada border, are some important floras in the Warner Range. In August 2005, my student crew, family, and I camped there to sample this region. Northeastern California is an amazing place. It is mostly high, dry plateaus of very young lavas, including some eruptions from the southernmost active volcanoes of the Cascade Range, Mount Lassen and Mount Shasta. Northeast California is wide-open country, with a small population supported mainly by ranching on sage-covered flats and by logging in the forests on the peaks. Due to the high elevation, the summers are not blazing hot, but the area is remote and rugged, with long drives from any one destination to another. We found a nice campground that was a part of a golf resort near Likely, California—a real surprise because the region is generally too dry to support anything but sagebrush, let alone a grassy golf course. Nevertheless, it was a good base for our excursions in the region, and we could come back to camp and showers at night. We spent a long day driving from Alturas, over the pass across the Warner Range on Highway 299, down into Cedarville, then along the east flank of the range, where the major plant localities were found. My colleague Jeff Myers of Western Oregon University, who has worked up these fossil localities for years, gave us directions. With guidance from the owner of the nearest ranch, Sara Gooch, we hiked up a long canyon road. We eventually located the rich plant-bearing locality in Granger Canyon, which is dated at 31.5 million years old. We got good magnetic samples from the limited exposures in this area, and the lab results were excellent (Prothero 2007b).

These floras in the Warner Range west of Cedarville, California, are the next step in deciphering the details of the Eocene–Oligocene floral transition (Myers 2003). In addition to the early Oligocene Granger locality (31.5 million years old), there are also late Eocene (34 to 35 million years old) fossil plants from the Badger's Nose and Steamboat localities. According to Jeff Myers (2003), the Steamboat floras give a mean annual temperature estimate of 17.1°C (62.8°F), the Badger's Nose floras grew at a temperature of 12.5°C (54.5°F), and the early Oligocene Granger floras experienced a mean annual temperature of 9.6°C (49.3°F), so these three floras in succession document a drop of about 7.5°C (13.5°F) over several million years. As I finished writing this book, the NSF came through with funding to do the research on this problem. So in the summer of 2009, I will be back up in the Warner Range to finish the sampling there and document the timing of this temperature drop much more precisely than we can right now.

Eel Ears, Turtle Shells, and Mammal Teeth

From our limited understanding of the Eocene–Oligocene transition in the late 1970s, we have come a long way. The North American land mammal sequence was once poorly dated, but now mammalian fossils can be dated to the nearest few thousand years or so. The North American terrestrial Eocene–Oligocene boundary was in the wrong place, at the Duchesnean–Chadronian boundary, but now we know that it occurs at the Chadronian–Orellan boundary. It had the wrong numerical age, once though to be 36 or 38 million years ago, but now we know it was 33.9 million years ago. The mammal fossils of this interval had been neglected for a generation, but now many different students are working on them, and taxonomic revisions of nearly all the important groups have been published (papers in Prothero and Emry 1996). Likewise, the understanding of land plant record has come a long way, with much more precise dating, resulting in a less extreme interpretation of how rapid and how severe the early Oligocene cooling was.

The study of the marine Eocene–Oligocene record has been reborn as well, with many new scientists looking at old and new fossils in many new ways. For example, my good friend Linda Ivany has applied a novel approach to the temperature problem. Among the thousands of beautiful fossils in the Gulf Coast Eocene–Oligocene record are the tiny ear stones (otoliths) of a shallow-water fish related to the conger eel (figure 7.6*A*). The stones were located in the eel's inner ear and helped it detect changes in its motion in the water as its head moved. They grew like trees, adding a ring of new material each year. Through a wonderful new technology, she can now slice open these tiny lumps of stone and then microsample them in increments of less than a millimeter so that she can get a lot of little bits of material from within every annual growth line in the fish's life (figure 7.6*B*). She then runs these tiny samples through a mass spectrometer, which can measure the isotopes of carbon and oxygen recorded by each ear bone of each eel (figure 7.6*C*). Linda has done this for hundreds of specimens spanning the entire Eocene and early Oligocene. Surprisingly, her results do not show as dramatic a temperature change as do the oxygen isotopes recorded in deep-sea foraminifera through most of the middle and late Eocene. However, the early Oligocene samples suggest that winter temperatures dropped at least 4°C (7°F)

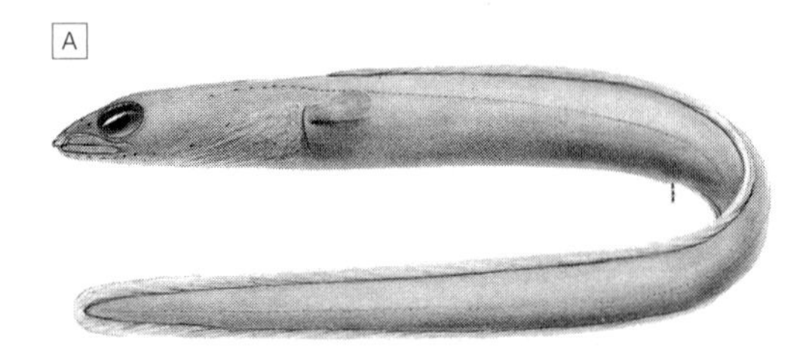

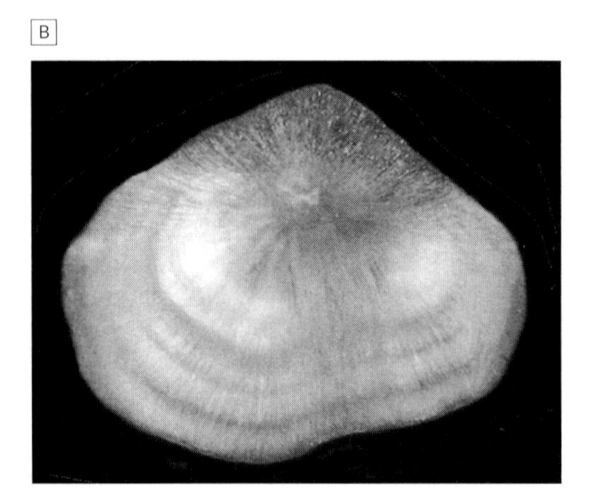

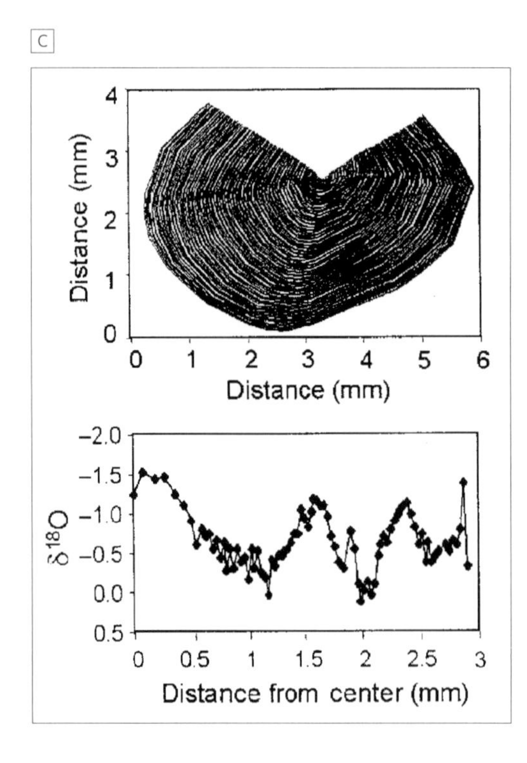

FIGURE 7.6 Isotopic analysis of an eel's ears: (*A*) the eel *Paraconger caudilimbatus*, related to the conger eel, a common bottom-dwelling fish in shallow waters; (*B*) photograph of the pea-size ear stone, or otolith, of *Paraconger*, a very common fossil in shallow marine sediments, showing the concentric growth lines formed during the eel's lifetime. The changes in temperature in the eel's seafloor environment can be deciphered by microsampling each tiny growth line and analyzing the isotopes of oxygen and carbon preserved in each layer. (*C*) Typical seasonal fluctuation of isotopes from a single eel otolith, showing several changes of season during its lifetime. (From Ivany, Lohmann, and Patterson 2003; images courtesy L. C. Ivany)

on average, a significant cooling. More important, they show a much greater range of temperature between winter and summer, so the key factor in the environmental changes of the early Oligocene is increased seasonality rather than decreased temperature (Ivany, Patterson, and Lohmann 2000; Ivany, Lohmann, and Patterson 2003).

Linda has also looked at a wide variety of phenomena, again using the technique of sampling tiny growth lines for their chemical isotopes. For example, she and Devin Buick sampled the growth lines of the clam *Cucullaea raea* from the Eocene La Meseta Formation of Seymour Island on the Antarctic Peninsula (Buick and Ivany 2004). They found that temperatures were much warmer than today (about 14°C [57°F], rather than below freezing as it is now) and that the clam shells grew very slowly and only during the Austral winters of total darkness. Surprisingly, they stopped growing entirely during the Austral summer when food was more abundant, probably because they were putting all of their energy and resources into reproduction. Due to their slow metabolism and growth rates, these clams were extremely long lived. Some had growth lines representing more than 100 years, so they lived longer than most humans do (figure 7.7A)! Linda and her colleagues studied other isotopes from shells in the same area (Ivany, Lohmann, et al. 2008). They found that mean annual temperatures dropped about 10°C (18°F) over the course of the middle Eocene, from about 15°C (59°F) in the early Eocene to only 5°C (41°F) in the latest Eocene. Although temperatures dropped rapidly at the end of the middle Eocene, there is no evidence of freezing on Seymour Island until the end of the Eocene (although Antarctica glaciers are known from elsewhere).

Ivany and her coauthors (2004) microsampled the growth lines of the thick-shelled early Eocene cockle shell *Venericardia* from the Hatchetigbee Formation of Alabama (see figure 6.5A). They found that temperatures in the early Eocene oceans of Alabama were warmer (26°C [79°F]) than today and much less seasonal as well. These data suggested that the conditions in the early Eocene of Alabama were much more like the tropics, with higher average temperatures and very little seasonal fluctuation.

Many more secrets are locked in the bones and teeth of fossils already stored in museum cabinets, and they only await someone with the interest and appropriate training to unlock them. For example, Eric Dewar at Suffolk University in Boston used the foundation of the well-dated fossils of the White River Group to examine details that my original work could not decipher. Tim Heaton and I had plotted the first and last occurrences of fossil species in an attempt to see whether they went extinct or originated in times of maximum climate change (Prothero and Heaton 1996). As I discussed in chapter 4, they did not show change at climatic events (see figure 4.3). But Dewar (2007) looked at more subtle possibility: maybe the species did not change in response to climate by speciating or

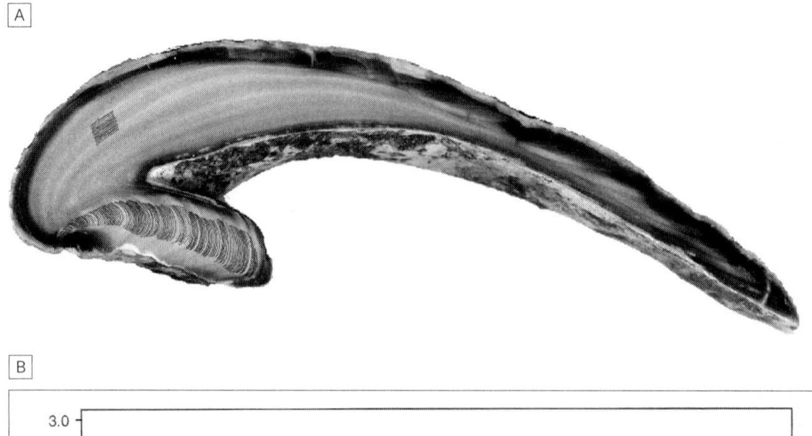

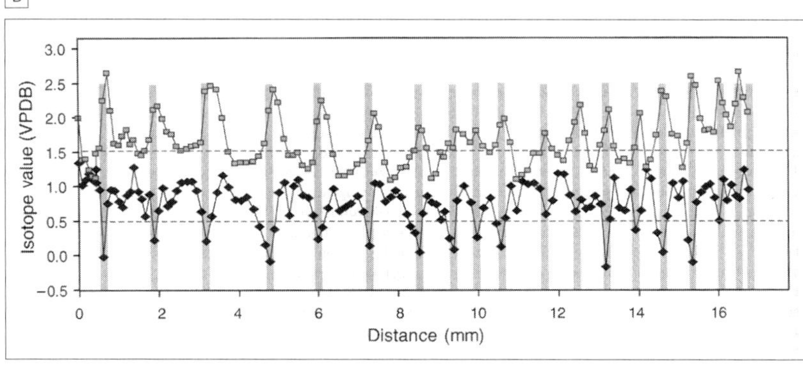

FIGURE 7.7 (*A*) Growth lines of the Eocene clam *Venericardia hatcheplata*; (*B*) pattern of isotopic change (VPDB = Vienna–Pee Dee Belemnite) in the growth lines of the Antarctic clam *Cucullaea raea* from the Eocene La Meseta Formation of Seymour Island, Antarctica, showing the many years of growth recorded by their shells. (From Buick and Ivany 2004)

evolving new shapes, but their behavior might have. He looked at the microscopic pits and scratches that accumulate on the enamel of teeth as they wear through an animal's lifetime. Some marks are clearly evidence of wear due to gritty grasses or similar vegetation, whereas other pits, gouges, and scratches might be due to eating hard seeds or nuts, or due to eating a gritty diet with lots of sand in it. He discovered that even though the shapes of the teeth of individual species of most hoofed mammals (horses, rhinos, oreodonts, camels, and so on) did not change through time, their diets did. He specifically found that the species shifted from a more browsing diet of leaves to a diet of more gritty grasses and shrubs (although there were still no extensive grasslands or savannas back then).

The carnivorous mammals all showed much evidence of bone breaking in their diets, with few that were strict meat eaters, in contrast to most of today's carnivores. This finding is consistent with what Blaire Van Valkenburgh of UCLA found in the 1980s (see Van Valkenburgh 1985). She plotted a three-dimensional hyperspace of modern carnivores based on dimensions of their feet, limbs, and overall body weight, and another hyperspace with axes comparing body weight to the size of the premolar teeth and the length of the bladelike teeth in the lower jaws (figure 7.8). These plots showed clear niche separation between the omnivorous bears, the pack-hunting dogs, the bone-crushing hyenas, and the specialized meat-slicing cats in today's modern ecosystems. But when she plotted the archaic hyaenodonts, the weasel-like early dogs, and the catlike nimravids of the White River fauna, they did not match the ecological niches occupied by living carnivores. Most of them clustered in the space of highly generalized carnivores (occupied by some dogs, cats, and mongooses today), but none showed the clear signatures of specialized running, as modern dogs and cats do. Nor were there any of the omnivorous types comparable to modern bears and raccoons or any strict meat eaters like modern cats (even though the nimravids look superficially like true cats). And the "hyenalike" *Hyaenodon* plotted nowhere near modern hyenas. The superficial comparisons to modern carnivores do not hold up when we try to force these ancient carnivorous mammal assemblages into modern categories.

There is evidence not only in the scratches and wear on the surface of the teeth, but also in the chemistry trapped in their enamel. Alessandro Zanazzi and colleagues (2007) looked at the isotopes of oxygen isotopes in the enamel of four different kinds of mammal teeth (the rhino *Subhyracodon*, the horse *Mesohippus*, the oreodont *Merycoidodon*, and the deerlike *Leptomeryx*) spanning the entire late Chadronian and early Orellan. They also looked at the isotopes recorded in turtle shells and in the carbonate nodules in the soils of the White River sediments of western Nebraska. They found that over 400,000 years of the Eocene–Oligocene transition, the mean annual temperature dropped about 8.2°C (14.8°F), and seasonality increased slightly. However, there was no evidence of increased aridity. This finding contrasts with the studies of Ivany and her colleagues (Ivany, Patterson, and Lohmann 2000; Ivany, Lohmann, and Patterson 2003) mentioned earlier, which did not find as much cooling, but found much more seasonality in the Gulf Coast marine record. Such a result is surprising because one would expect that the midcontinent terrestrial record would be much more seasonal than the buffered, warmer environment of the Gulf of Mexico.

Zanazzi and colleagues' (2007) results must be viewed with caution, however. Their samples come from sections in western Nebraska that do not have detailed high resolution across the Eocene–Oligocene transition. In particular, they are missing a detailed record of the earliest Orellan *Hypertragulus calcaratus* Interval Zone (Prothero and Whittlesey 1998). This interval is only well preserved in sections near Douglas and Lusk in eastern Wyoming. Until someone analyzes samples from this interval in eastern Wyoming, we will not be able to tell how abrupt and rapid the change really was.

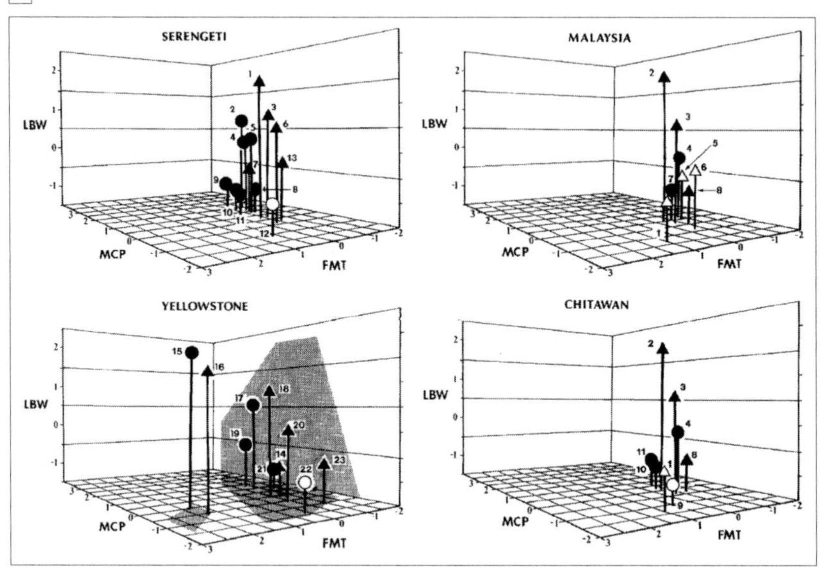

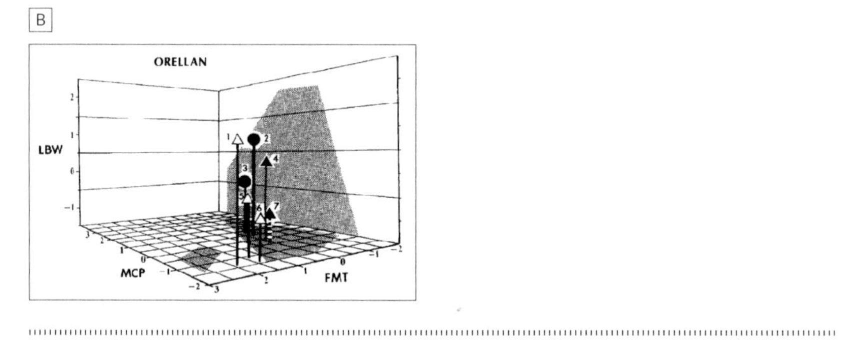

FIGURE 7.8 Three-dimensional plots of carnivorous mammals showing clustering by ecological specializations. The vertical axis is body weight (LBW), and the horizontal axes are the ratios of wrist bones to finger bones (MCP) and the ratio of the thighbone to the foot bones (FMT). Arboreal carnivores are shown by open triangles; scansorial (limited climbing) animals, by solid triangles; terrestrial carnivores, by solid circles; and burrowers, by open circles. (*A*) The Serengeti carnivores occupy a smaller variety of ecological niches than do the Yellowstone carnivores. Species are: *1*, lion; *2*, spotted hyena; *3*, cheetah; *4*, wild dog; *5*, striped hyena; *6*, leopard; *7*, serval; *8–10*, jackals; *11*, civet; *12*, ratel; *13*, caracal; *14*, bobcat; *15*, grizzly bear; *16*, black bear; *17*, wolf; *18*, puma; *19*, coyote; *20*, wolverine; *21*, red fox; *22*, badger; *23*, lynx. In the Malaysian and Chitawan predator plots, an even smaller range of locomotor types are present. Species are: *1*, binturong; *2*, tiger; *3*, leopard; *4*, dhole; *5*, Temminck's cat; *6*, clouded leopard; *7*, civet; *8*, fishing cat; *9*, ratel; *10*, jackal; *11*, civet. (*B*) In the early Oligocene of the Big Badlands, an even smaller range of locomotor types is represented, and some have no modern analogues. Species are: *1* and *6*, *Hoplophoneus* (saber-toothed nimravid); *2* and *3*, *Hyaenodon*; *4* and *7*, *Daphoenus* (beardog); *5*, *Dinictis* (dirk-toothed nimravid). (After Van Valkenburgh 1985, by permission of the Paleontological Society)

Many other tools are now being applied to the question from a wide range of scientific fields. For example, Nathan Sheldon and Greg Retallack (2004) looked at the ancient soil horizons preserved across the Eocene–Oligocene transition in Oregon, Montana, and Nebraska. They used two common tools that indicate climate change: the depth of the level in the soil where calcite is precipitated and the degree of chemical weathering. The depth of calcic horizon is an important measure because in dry climates it is very shallow, and thick layers of calcite known as caliche can be precipitated. In wetter climates, leaching from rainwater dissolves the shallow carbonate, and it precipitates at a much deeper level in the soil. Sheldon and Retallack found a long-term rather than sudden shift in temperature, with a gradual cooling throughout the transition. They also found a gradually increased trend in aridity, but no sudden drying event at the Eocene–Oligocene boundary. This finding is in agreement with the data from reptiles compiled by Howard Hutchison (1982), who argued that the early Orellan was marked by increased aridity because there was a significant increase in the number of dry-land tortoises and a disappearance of pond turtles and crocodilians. It also agrees with the sedimentological evidence described by Emmett Evanoff, myself, and Rob Lander (1992) from Douglas, Wyoming, where you can see sediments representing floodplains being replaced by wind-blown dune deposits in the early Orellan. But it is in striking contrast to the results suggested by Zanazzi and his coauthors (2007), who found no evidence of increased aridity in their isotopic results from mammal teeth, turtle shells, and soil carbonate.

Clearly, we have moved on to an interesting new phase of research that has yielded some answers but also some puzzles as well. None of it would have been possible without the foundation of fossil collections and stratigraphy pioneered by Morris Skinner in the 1940s and 1950s, which many of us completed and published in the 1980s and 1990s. With this body of work, any scientist with the proper training and interest can look for new solutions to these puzzles and give us yet more interesting data to think about. And maybe, if we're lucky, enough of these ideas and data will converge, and we'll begin to see the general truth behind the confusing cloud of so many different studies of different phenomena. As scientists, this convergence is what we hope to achieve—but even if we don't, the journey is fun and worth the effort in and of itself.

Further Reading

Graham, A. 1999. *Late Cretaceous and Cenozoic History of North American Vegetation*. Washington, D.C.: Smithsonian Institution Press.

MacGinitie, H. D. 1953. *Fossil Plants of the Florissant Beds, Colorado.* Carnegie Institute of Washington Publication, vol. 599. Washington, D.C.: Carnegie Institute.

Meyer, H. W. 2003. *The Fossils of Florissant.* Washington, D.C.: Smithsonian Institution Press.

Myers, J. 2003. Terrestrial Eocene–Oligocene vegetation and climate in the Pacific Northwest. In D. R. Prothero, L. C. Ivany, and E. A. Nesbitt, eds., *From Greenhouse to Icehouse: The Marine Eocene–Oligocene Transition*, 171–188. New York: Columbia University Press.

Prothero, D. R. 2006. *After the Dinosaurs: The Age of Mammals.* Bloomington: Indiana University Press.

Retallack, G. J., W. N. Orr, D. R. Prothero, R. A. Duncan, P. R. Kester, and C. P. Ambers. 2004. Eocene–Oligocene extinctions and paleoclimatic change near Eugene, Oregon. *Geological Society of America Bulletin* 116:817–839.

Wolfe, J. A. 1971. Tertiary climatic fluctuations and methods of analysis of Tertiary floras. *Palaeogeography, Palaeoclimatology, Palaeoecology* 9:27–57.

Wolfe, J. A. 1978. A paleobotanical interpretation of Tertiary climates in the Northern Hemisphere. *American Scientist* 66:694–703.

The *Glomar Challenger*, the oceanographic research vessel that drilled thousands of cores all over the world's oceans from 1968 to 1983 and made an amazing number of important geological discoveries. (Photograph courtesy Oceanic Drilling Program)

8 | From Greenhouse to Icehouse

> And now there came both mist and snow
> And it grew wondrous cold
> And ice, mast-high, came floating by
> As green as emerald.
>
> —SAMUEL TAYLOR COLERIDGE, *THE RIME OF THE ANCIENT MARINER*

From the Ashes of Disaster . . .

The general public hears only about the successful experiments in science. What is usually not reported is the fact that for every scientific success, there may be numerous failures, false leads, and blind alleys. Most people would find this imbalance discouraging, but scientists learn early in their careers that they should expect a number of failed experiments, but also that these failures can lead them to better ideas. As philosophers of science pointed out long ago, science is about testing and falsifying hypotheses. No number of positive or consistent observations can ever prove a statement true (e.g., "all swans are white"), but a single contradictory observation (e.g., the Australian black swan) can easily prove the statement false. Likewise, every failed experiment points the scientist toward a new direction or new hypothesis or new experiment, which may eventually prove fruitful. Science is a process of trial and error, and scientists need patience, persistence, and determination to reach good results after many letdowns.

Nearly every field in science can point to examples of this process. As I described earlier in the book, not every locality where I have done paleomagnetic sampling produced good results. The data from those unsuccessful studies are sitting in my file cabinets and the hard drive of the lab computer, but I won't bother working on them further or try to report them in a publication. I occasionally mention in print that a particular area (such as the Chadron Formation in the Big Badlands or the Titus Canyon Formation near Death Valley) produced no good paleomagnetic results, but further discussion about them is for the most part not worth writing up.

Likewise, looking for vertebrate fossils is usually a frustrating and unsuccessful exercise. Most paleontologists must spend days or weeks in a field to find anything, and several field seasons can sometimes go by with no worthwhile results. Malcolm McKenna spent several years collecting his dissertation area at Four Mile Creek in northwestern Colorado before finding good specimens. Louis and Mary Leakey spent decades collecting in Olduvai Gorge, Tanzania, and found plenty of Pliocene–Pleistocene pigs, antelopes, and elephants, but not a single human fossil until their remarkable find of "*Zinjanthropus*" (now *Paranthropus*) *boisei* in 1959. But if paleontologists were not so determined and dedicated, there would be no fossils in museums for us to study.

Similarly, the history of marine geology, paleoclimatology, and oceanography would have never been the same were it not for a little bit of luck and persistence in the face of failure. Back in the early 1950s, the famous geophysicist Walter Munk argued that we needed to develop a technology to drill down to the mantle. Many considered this goal comparable to our effort to put rockets in space. Because the continent crust is typically more than 150 to 200 kilometers (93 to 124 miles) thick, but the oceanic crust is only about 10 kilometers (6.2 miles) thick, it made sense to fit a drilling rig on a ship and try to drill through the oceanic crust. From this idea developed Project Mohole, an NSF-funded effort to build a drilling ship and technology that could drill down to the Mohorovicic discontinuity, or "Moho," the seismic boundary between the crust and the mantle (the "hole to the Moho" or "Mohole"). The first drilling took place in March and April 1961, at which time the ship managed to drill down to 183 meters (601 feet) below the sea bottom in 3,566 meters (11,700 feet) of water off the coast of Mexico. Most of this section was made up of Miocene sediments that were easy to drill, but the basement was oceanic crust lavas, which made drilling much more difficult. Consequently, they never got even close to penetrating 10 kilometers (6 miles) of oceanic crust. After these initial experiments, the enthusiasm for Project Mohole fizzled out, and the project eventually ended by 1966 as the host organization disbanded in a dispute with NSF over its control.

Come the Roses of Success

Project Mohole failed in its initial plans to drill to the mantle, but it was a smashing success in a different way. The early experiments showed that although the technology of the 1960s couldn't drill too far in oceanic crust lavas, it did a beautiful job drilling the overlying sediments. As Project Mohole was falling apart, other scientists saw an opportunity: drilling the soft sediments of the seafloor was a much more important goal. They (correctly) guessed that the seafloor sediment records might yield important information about how oceans and climates had changed over millions of years. Because oceanic sediments rain down continuously from the surface waters, the record in the deep sea would be much more continuous and complete than any record obtained from land sections.

In June 1966, the Deep-Sea Drilling Project (DSDP) was formed, and a new ship, the *Glomar Challenger*, was commissioned (see the photograph that opens this chapter). Built by the Levingston Shipbuilding Company of Orange, Texas, in 1967 and 1968, it was run by the Global Marine cooperation (hence "Glomar"), a consortium of oil companies and academic agencies, and partially funded by the NSF. The *Glomar Challenger* was named after the famous British sailing ship H.M.S. *Challenger*, which sailed around the world from 1872 to 1876 and provided the very first systematic data from the world's oceans. However, the *Glomar Challenger* was much larger: more than 121 meters (375 feet) in length, with a draft of 6 meters (almost 20 feet). On top of the middle deck was a 43-meter (130-foot) drilling derrick, which could handle 6,860 meters (22,507 feet) of drilling pipe, drill in water depths of 6,000 meters (more than 19,000 feet), and penetrate the seafloor down to 762 meters (more than 2,400 feet). These technologies were developed originally by Project Mohole and by several oil companies for oceanic drilling, but the *Glomar Challenger* had additional technological advances. One of the most significant was a series of thruster propellers on bow and stern that could keep the ship in position over a given spot on the seafloor for days at a time, allowing it to drill in one spot continuously. The ship maneuvered by placing a sonar beacon on the seafloor, so it could track its position and correct when it drifted off the drill site. In 1970, the marine engineers and scientists developed an additional bit of amazing technology: a series of small thrusters on the tip of the drill string. These thrusters allowed the ship to pull up its old drill string, replace the worn-out drill bit, and put thousands of feet of pipe back in the ocean. The thrusters at the tip would then guide the drill bit, home in on a sonar beacon placed in the seafloor, reenter the original hole, and drill much deeper than ever before. Some people have compared the process of deep-sea drilling to trying to manipulate a single extremely long strand of wet spaghetti from a third-floor balcony and making it touch exactly the same spot on the ground again and again (figure 8.1). Yet this amazing feat was done routinely dozens of times on every voyage of the *Glomar Challenger* and is still done today.

The *Glomar Challenger* was stocked with provisions to stay at sea for up to 90 days. It had a crew of sailors, roughnecks recruited from the oil companies to handle the manual labor of the drilling, and a revolving shipboard scientific crew responsible for curating and analyzing the deep-sea cores as they came up

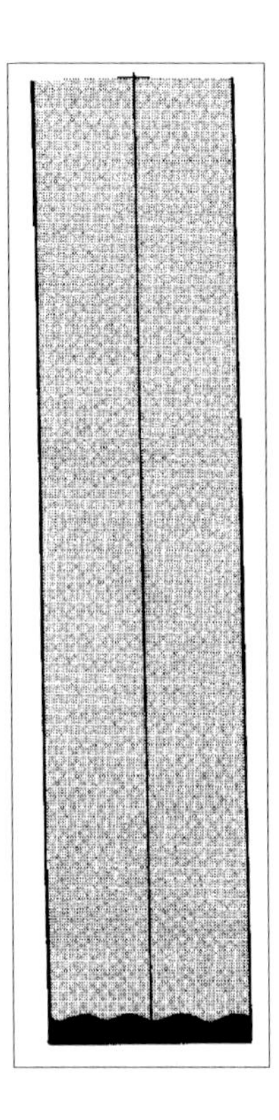

FIGURE 8.1 The scale of the *Glomar Challenger*'s drill string compared with the size of the ship in 5,500 meters (18,000 feet) of water. In actuality, the drill string is so thin at this scale that it would not even be visible. (Modified from Kennett 1981)

on deck. Its first shakedown cruise (Leg 1) in 1968 drilled in the Gulf of Mexico and found salt domes that proved very valuable to the oil companies who'd helped fund the project. The first purely scientific cruise, Leg 3 in 1969, systematically drilled sites at varying distances away from the Mid-Atlantic Ridge in the South Atlantic. These sites showed that the sediments and crust get older away from the ridge, confirming the predictions made by the newly proposed theory of seafloor spreading and plate tectonics. Leg 13 in 1970 drilled a series of holes in the Mediterranean and found evidence that the Mediterranean Sea had dried up completely about 6 to 5 million years ago, producing a huge desert basin like a gigantic Dead Sea or Death Valley several miles deep, which was then flooded at the beginning of the Pliocene when the Gibraltar dam was breached. By the mid-1970s, James Hays, John Imbrie, and Nicholas Shackleton (1975) used data from many Pleistocene sedimentary cores to demonstrate that the Ice Age cycles were controlled by variations in Earth's orbit around the Sun.

As a graduate student at Columbia and the Lamont-Doherty Geological Observatory (now Lamont-Doherty Earth Observatory) in the late 1970s and early 1980s, I had the opportunity to see and hear firsthand about the DSDP discoveries.

Much of the effort for paleooceanography and marine geology was directed out of Lamont, which had two oceanographic research ships of its own, taking deep-sea cores in oceans around the world. These cores were stored in a huge core lab at Lamont, along with many of the DSDP cores from the *Glomar Challenger*. Many of the scientists who had been on DSDP cruises taught me marine geology and micropaleontology in my Lamont classes. My marine geology professor, Bill Ryan, had been a pioneer in developing marine technology and was one of the chief scientists on the 1970 cruise that discovered the Mediterranean had been a desert, as well as on many other legs of the DSDP. Jim Hays, an expert on late Cenozoic microfossils called radiolaria, had developed the first biostratigraphy of Antarctic radiolaria based on his DSDP work and eventually helped discover the pacemaker of the ice ages. He not only taught me about radiolaria, but got me involved in micropaleontology research with his own grad student Dave Lazarus. I spent many months in the core lab on our joint project on evolutionary change in radiolaria, measuring hundreds of radiolaria on dozens of slides. Jim Hays was also on my dissertation committee. I learned about diatoms from Lloyd Burckle, a DSDP veteran and a world expert on the group, and about foraminifera from Tsunemasa Saito, also a scientist on the DSDP. And my paleomagnetics professor, Neil Opdyke, not only was one of the pioneers of paleomagnetism, but also developed the first technology for doing shipboard paleomagnetism on deep-sea cores, which essentially started the field of magnetic stratigraphy.

Over the years, I heard story after story about shipboard life on the *Glomar Challenger*, which ranged from the long grind of hard work when cores were coming in one after another to long periods of boredom when cruising between sites, especially before the first cores were retrieved. There was shipboard camaraderie in the scientific party, and many pranks and gags filled up the time. And, of course, each time the ship crossed the equator, anyone who had never before changed hemispheres on the open ocean ("Slimy Pollywogs") got to be part of the centuries-old maritime tradition of the "Crossing the Line" ceremony. The Pollywogs would appear before King Neptune and his court (Davy Jones and Amphitrite) and submit to gentle hazing by those who had been across the equator previously (the "Trusty Shellbacks").

Many DSDP legs over the next decade provided more and more evidence about the history of the oceans and ancient climates as well. In short, the DSDP has been one of the most important scientific projects of the entire history of geology, teaching us more about oceans and climate than we learned in the entire previous century. In terms of important dividends of both scientific and economic value, it has more than justified the millions of federal dollars spent on it. And if it weren't for the failure of Project Mohole, it would never have happened.

By the late 1970s, the *Glomar Challenger* was becoming outdated and wearing out, and a new vessel was needed. After 96 legs of its voyage around the world many times over, 375,632 nautical miles, and 19,119 cores recovered representing more than 170,043 meters of core, the *Glomar Challenger* was finally taken out of commission in November 1983 and later scrapped, which in itself is a sad thought: an amazing piece of scientific history reduced to scrap metal. It's comparable to scrapping *Old Ironsides* or the U.S.S. *Missouri* (the battleship on whose deck Japan surrendered in 1945) or historic planes such as the Wright *Flyer* or Lindbergh's *Spirit of St. Louis* (both now in the Smithsonian).

Nevertheless, in 1985 a new ship, the *JOIDES Resolution*, was commissioned for the Oceanic Drilling Program (ODP), the successor to DSDP. It was named after the H.M.S. *Resolution*, the ship that Captain James Cook used to explore the Pacific and sail around the world more than 200 years ago. It was much larger than the old *Glomar Challenger*. Its derrick was 61.5 meters (202 feet) above the water line, and it was 143 meters (469 feet) long and 21 meters (68.9 feet) wide. During a leg, the crew positioned the ship over the drill site using 12 computer-controlled thrusters as well as the main propulsion system. The drilling apparatus could suspend as much as 9,150 meters (30,020 feet) of drill pipe to an ocean depth as great as 8,235 meters (27,018 feet). The 50 scientists and 65 crew members lived in the forward section of the ship. On the aft part of the deck was a "lab stack," a seven-story building with more than 12,000 feet of space that housed the labs and the cores. Some of the labs were used for cold storage of the cores after they left the deck. Other decks housed computers and micropaleontology and sedimentology labs for describing the details of each core as it reached the lab. Another deck had labs to do paleomagnetic analysis of the core on board the ship. Even though the ship would be at sea for only two to three months before returning to a port of call for provisions, the work on the ship went on 24 hours a day, seven days a week. After each two- to three-month cruise leg, the scientists rotated off and returned home, and a new group of scientists boarded the ship for another leg planned for a different part of the ocean and different research goals.

ODP and the *JOIDES Resolution* completed another 110 legs around the world's oceans before the program ended in 2003. Since 2006, the Integrated Ocean Drilling Program (IODP) has replaced the ODP and is supported not only by the NSF, but also by Japanese and European academic and scientific institutions. It has two ships: the refitted *JOIDES Resolution* and a Japanese ship, the *Chikyu* (Earth). The latter is nicknamed the *Godzilla Maru* because it is the largest research vessel ever built and can drill deeper than any ship ever floated. Ironically, it has technology that will allow it to drill deep in the oceanic crust (which it had already done by 2005), and the plan is eventually to drill down to the mantle by 2012. Thus, we have come full circle. The failure of Project Mohole to reach the mantle led to drilling oceanic sediments with DSDP and ODP instead, producing a wealth of unexpected scientific studies. As a consequence, the drilling technology advanced, and now, 50 years after Project Mohole, the *Chikyu* is on the verge of accomplishing the original goal of reaching the mantle.

Antarctic Surprise

Most of the DSDP and ODP legs were important and produced many significant scientific discoveries, but some (as mentioned earlier) generated data that transformed all of geology. Two of the most important early voyages were DSDP Leg 21 (November 1971 to January 1972, from Fiji to Darwin, Australia, drilling Holes 203 to 210) and Leg 29 (March to April 1973 from Lyttleton to Wellington, New Zealand, drilling Holes 275 to 284). Both legs were among the first to drill the southwestern Pacific Ocean and the Southern Ocean around New Zealand and eastern Australia, with some sites only a few degrees of longitude from the Antarctic coastline. Most of the sites from Leg 21 were to the north of New Zealand along Lord Howe Rise (a shallow submarine ridge running northwest of New Zealand), in the Coral Sea south of New Guinea, or in the South Fiji Sea. Most of the sites from Leg 29 were around Tasmania or on the Campbell Plateau or Macquarie Ridge, shallower features on the deep-sea bottom that are due south of New Zealand. In both cruises, long cores were obtained that showed the first deep-marine records of the Eocene and Oligocene in the Southern Ocean. Taken together, these cores provided the first evidence of deep-marine cooling in the Oligocene in the Antarctic. This finding led the marine geologists and paleontologists on these cruises (Kennett, Burns, et al.1972; Kennett, Houtz, et al. 1975; Kennett 1977, 1981) to suggest that the Oligocene cooling was due to the opening of the deep-water passage between East Antarctica and Tasmania, and to the development of the circum-Antarctic current (discussed more fully later in this chapter).

We know quite a bit about the stages of the refrigeration of Antarctica through the Eocene. By the beginning of the middle Eocene, 49 million years ago, there were small glaciers on the Antarctic Peninsula and West Antarctic (Birkenmajer 1987; Birkenmajer et al. 2005). By the late middle Eocene, icebergs were melting in the Pacific sector of the Southern Ocean and dropping ice-rafted sediment, which has been retrieved from deep-sea cores (Wei 1989). Yet Antarctica had not yet frozen over completely even then, as indicated by the cool-temperate forests of southern beeches, araucarias (monkey puzzle trees or Norfolk Island pines), podocarps, and ferns over much of the continent in the middle and late Eocene (Kemp 1975; Case 1988; Mohr 1990). Only after the great expansion of the early Oligocene ice sheets at 33 million years ago did the Southern Hemisphere switch to full "icehouse" mode.

So what drove this gradual cooling of the Southern Hemisphere? There is no shortage of ideas and models, but many unanswered questions. The major explanations fall into several categories, with the main ones focusing on regional cooling due to changing oceanic circulation or global cooling due to falling atmospheric carbon dioxide levels.

Gateways in the Oceans

DSDP Legs 21 and 29 near New Zealand, Tasmania, and Antarctica provided the first evidence for the Eocene–Oligocene transition in the Southern Ocean. The most striking pattern was changes in the plankton, when silica-rich marine algae (diatoms) became important in the Southern Ocean, suggesting an upwelling of silica along the Antarctic margin. In addition, the chemistry of the planktonic foraminifera shells suggested both cooling and changes in oceanic productivity during the early Oligocene. Jim Kennett, Nick Shackleton, and others from the DSDP Legs 21 and 29 (Kennett, Burns, et al., 1972; Kennett, Houtz, et al. 1975; Kennett 1977; see also Berggren and Hollister 1977) argued that the Oligocene cooling was due to the onset of the circum-Antarctic current, also known as the Antarctic Circumpolar Current (ACC). Today this current circulates around Antarctica in an easterly direction, or clockwise if one is looking down on the South Pole (figure 8.2). It is the fastest and most voluminous of all currents in the ocean, traveling at 25 centimeters (10 inches) per second and with a volume of 233 million cubic meters (305 cubic yards) per second, or more than 1,000 times the flow of the Amazon (Callahan 1971)! As this current moves around Antarctica, it acts like a refrigerator door that helps lock the cold over the Antarctic continent and prevents warm temperate and tropical waters from reaching the Antarctic coastline. The upwelling of cold waters in rather complex patterns brings nutrients to the surface, producing huge blooms of diatoms in the plankton, which in turn nourish the plankton feeders of the Southern Ocean, such as the krill and the great baleen whales that now flourish there.

The Antarctic Bottom Water (AABW) also traces its origin to the chilling of the Antarctic. This cold but oxygen-rich current flows along the very bottom of the ocean floor up from the Antarctic all the way to the northwestern Atlantic and flows into the deeper parts of the Indian and Pacific oceans. It is the driver of the modern *thermohaline circulation system*, also called the *meridional overturning system*, which is the system of currents ventilating the deep oceans today. The AABW makes up 59 percent of the water in the oceans, so it is very influential in global climate (Warren 1971). Before the late Eocene, there was no ACC, so the

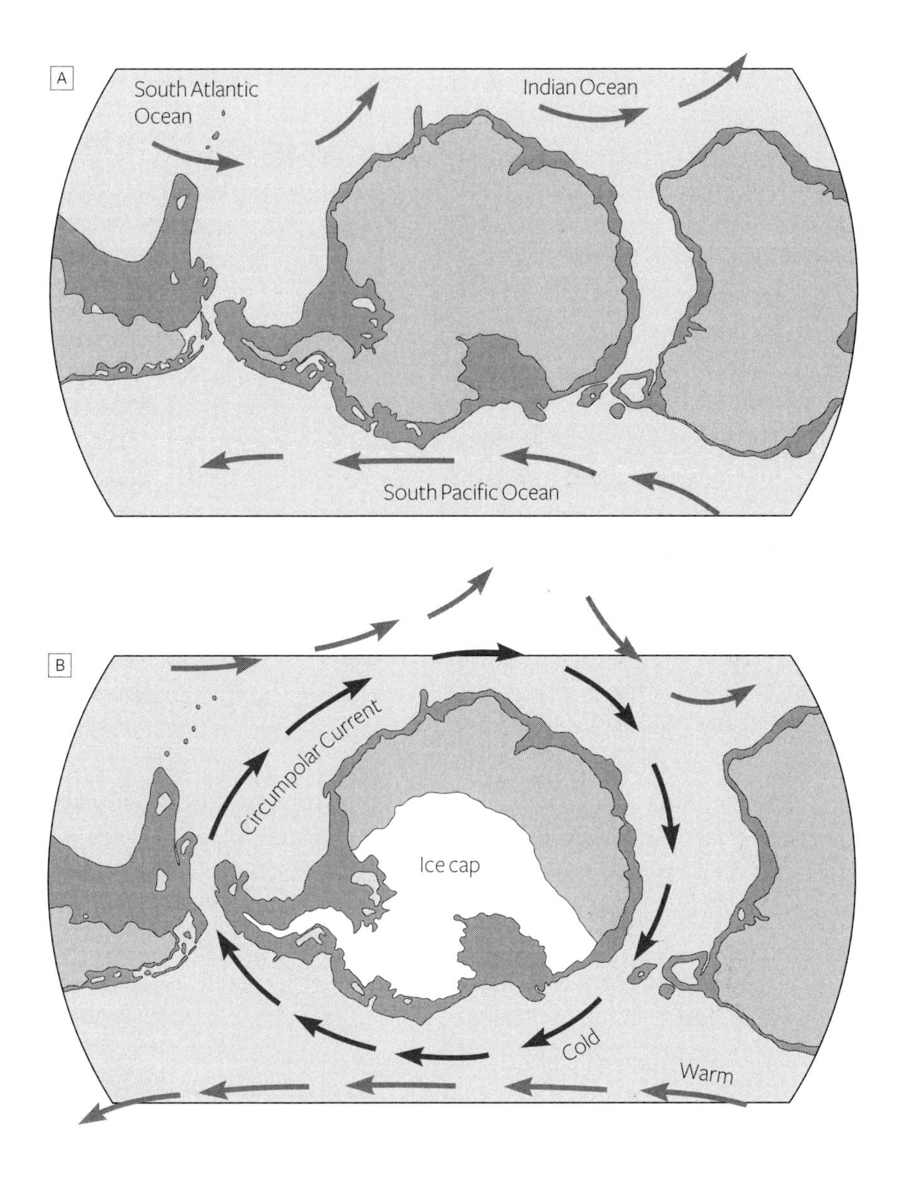

FIGURE 8.2 The Antarctic Circumpolar Current: (*A*) during the Eocene, Australia and South America blocked the ACC, allowing temperate and tropical waters to mix with polar waters; (*B*) the ACC as it flows today.

tropical and temperate waters may have mixed with the polar waters, warming the poles, and there was thus much less of a gradient in temperatures between the poles and the equator (see figure 8.2).

When did the modern current system originate? According to some plate tectonic reconstructions, Antarctica had been pulling away from Australia since the Late Cretaceous, allowing the first shallow marine waters to flow between them by the middle Eocene (see figure 8.2). Deep-sea cores from just south of New Zealand have long suggested that a blast of cold deep water passed through the Antarctic–Australia gap (figure 8.3) in the earliest Oligocene (Murphy and Kennett 1986; Kamp, Waghorn, and Nelson 1990; Exon et al. 2002). According to the long-accepted theory, this early Oligocene current coincides nicely with the growth of the first major continental ice sheets on Antarctica, which during the Eocene had been cool and vegetated, not covered by an ice cap.

To complete the ACC, however, there needed to be a gap between South America and the Antarctic Peninsula (the Straits of Magellan or Drake's Passage) to allow cold water to pass from the Pacific side to the Atlantic side of the Southern Ocean. Estimates originally placed the opening of Drake's Passage in the late Oligocene or even early Miocene, but each of those dates is too late to contribute to the Eocene–Oligocene cooling (Barker and Burrell 1977, 1982; Sclater et al. 1986). In the past decade, the data (Diester-Haass and Zahn 1996) seemed to suggest that the passage was open in the early Oligocene. If so, then the ACC could be completed at that time, and the current would have frozen the Antarctic and produced its first ice caps. Roy Livermore and his colleagues (2007) argue that the Drake Passage may have been open as early as the middle Eocene. However, Mitchell Lyle and coauthors (2007) looked at the geochemistry of cores from the South Pacific and concluded that the ACC was not fully developed until the late Oligocene, but not as late as the Oligocene–Miocene boundary, when there is a big cold pulse. Thus, the problem keeps getting more interesting and complicated as the answers change with each new source of data. Unfortunately, with the end of the ODP and the changing goals of oceanic drilling, few cruises are taken in this critical region anymore, so it is unlikely that much new data will be generated from oceanic drilling to test these ideas.

The story has yet another twist. The deep, cold bottom water from the Arctic Ocean flows out through the Norwegian-Greenland Sea and into the North Atlantic and produces another water mass known as the North Atlantic Deep Water (NADW), which flows over the AABW in the western Atlantic. Several studies of deep-sea cores have suggested that the NADW originated in the early Oligocene, further helping to cool the planet and drive it into the icehouse state (Miller and Fairbanks 1983; Miller and Tucholke 1983; Miller and Thomas 1985; Miller 1992; Davies et al. 2001). Others argue that the NADW originated much later (Poore et al. 2006). If the first group of studies is right, and both the NADW and ACC

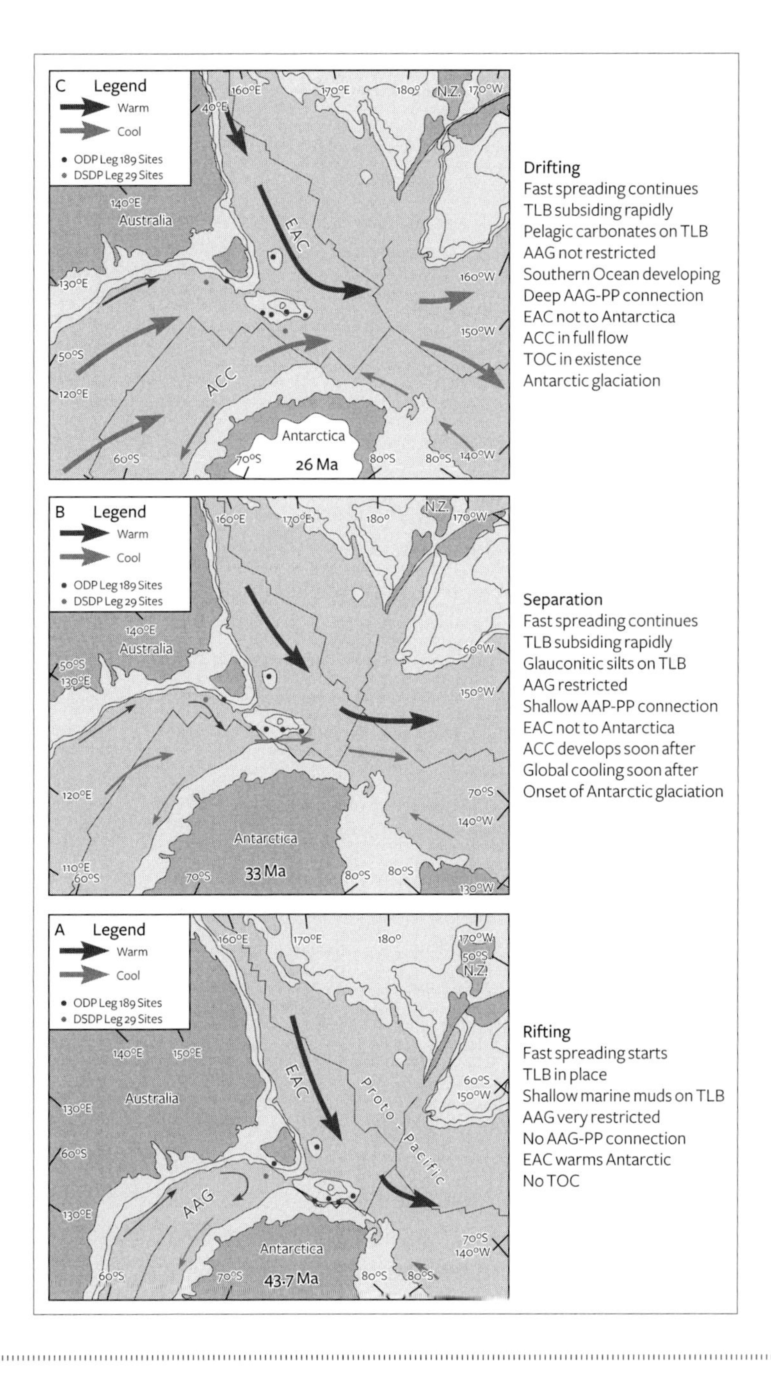

FIGURE 8.3 The reconstruction of oceanic circulation patterns advocated by Neville Exon and colleagues (2002) (modified from Murphy and Kennett 1986), which suggests a deep-water flow south of Tasmania by the early Oligocene. In this model, the East Antarctic Current flows counterclockwise out of the north, providing the cold water detected in cores south of New Zealand.

originated in the early Oligocene, there would have been powerful forces at work to chill Antarctica, form the first polar ice caps since the Permian, and drive the planet into an icehouse state.

This idea of northern refrigeration in the Eocene was recently confirmed by the discovery of glacial dropstones that rafted from icebergs off Greenland in the middle Eocene, about 44 to 38 million years ago (Moran et al. 2006; Eldrett et al. 2007; Tripati, Eagle, et al. 2008) and in the central Arctic Ocean (Moran et al. 2006). These data suggest that Greenland had significant ice caps in the late Eocene and Oligocene, which goes against the conventional wisdom that the Arctic was warm and ice free until the modern Arctic ice cap arose in the late Miocene. Stefan Schouten and colleagues (2008) recently described evidence from a sediment core in the Greenland Basin that shows the temperature dropped in the Northern Hemisphere from about 15°C (59°F) in the middle and late Eocene and declined about 5°C (41°F) across the Eocene–Oligocene boundary, consistent with the idea of a Greenland ice cap. The temperature drop suggested for the southern high latitudes is about 10°C (18°F) at the end of the Eocene, even more extreme than for the Northern Hemisphere (Dutton, Lohmann, and Zinsmeister 2002; Poole, Cantrill, and Utescher 2005). By contrast, Paul Pearson and coworkers (Pearson, van Dongen, et al. 2007; Pearson, McMillan, et al. 2008) showed that tropical marine waters remained unchanged in temperature through the Eocene–Oligocene transition.

In addition, other researchers (Coxall et al. 2005; Tripati, Backman, et al. 2005) have documented how the oceans became much cooler more rapidly in the late Eocene and early Oligocene than previously suspected. The temperature drop took place in two steps over less than 80,000 years. According to these authors, these changes are too large to be explained by Antarctic glaciation alone, but also require some sort of Arctic–Greenland glaciation as well.

Some scientists, however, doubt that oceanic currents can explain the magnitude of the cooling in the late Eocene. For them, the intuitively attractive idea of the thermal isolation of the Antarctic driving the icehouse—an idea that dates back to Leg 21 (Kennett, Burns, et al. 1972)—has been slipping out of favor. According to these atmospheric and oceanic models, Antarctic ice sheets first formed when concentrations of greenhouse gases dropped in the early Oligocene (Deconto and Pollard 2003; Pagani et al. 2005). Robert DeConto and David Pollard (2003) and Matthew Huber and his colleagues (2004) argued that changes in oceanic heat transport due to the opening of the Tasmanian Gateway between Australia and Antarctica was insufficient to explain the cooling of Antarctica. Matthew Huber and Doron Nof (2006) used atmosphere–ocean climate models to simulate both the effects of changes in oceanic circulation and changes in carbon dioxide in the atmosphere. They found that oceanic currents were insufficient to produce an ice sheet, even considering such positive feedback as the increased heat loss to outer space due to the presence of ice and the increased absorption of carbon dioxide in the oceans due to cooling. Just dropping the carbon dioxide levels significantly led to the cooling of the Antarctic and Southern Ocean. Other modeling results also suggest that the change in meridional heat transport due to the onset of the ACC was insignificant (Huber and Sloan 2001; Huber, Sloan, and Shellito 2003; Huber et al. 2004; Huber and Nof 2006). For instance, Kennett (1977) assumed that relatively warm surface waters flowed southward along Australia, reaching the Antarctic continent, keeping that region warm. Recent biogeographic data and climatic modeling, however, show that such southward circulation probably did not exist in the Eocene. Instead, more recent coring data described by Catherine Stickley and colleagues (2004) show that a shallower Proto–Leeuwin Current and Proto–East Antarctic Current (figure 8.4) passed south of Tasmania, but not that a large, cold, deep-water current was traveling east between Antarctica and Australia, as others had proposed (Murphy and Kennett 1986; Exon et al. 2002) (see figure 8.3). The earlier data from coring in the 1970s that suggested this evidence of a cold current may instead be explained by the northward deflection of the cold Tasman Current up from Antarctica and past New Zealand, not by a cold current passing south of Australia (see figure 8.4).

More recent coring data found a counterclockwise gyre in the South Pacific rather than a clockwise circulation along the Antarctic coast (Huber et al. 2004; Stickley et al. 2004). Bilal Haq (1981) showed such a gyre in his simplified figures of oceanic circulation. In summary, there has been declining acceptance of the idea that the cooling of Antarctica was caused mainly by oceanic circulation, and the roles of greenhouse gases and weathering feedbacks (discussed in the next section) are seen as more significant forcing factors.

The Mountains Rise, the Oceans Widen

In contrast to oceanographic models for Eocene–Oligocene climate change, a number of authors (Ruddiman and Kutzbach 1991; Raymo and Ruddiman 1992) have argued that the rapid uplift of new mountain ranges (especially the Himalayas) during the Cenozoic would also have increased the rate of crustal weathering, removing carbon dioxide from the atmosphere as it combined with newly formed minerals in the soils. However, this model doesn't work too well for the early Cenozoic because the uplift of these ranges (such as the Himalayas or Rockies) occurred mostly in the past 15 million years (Rea 1992), too late (in the middle to late Miocene) to have much to do with Eocene or Oligocene

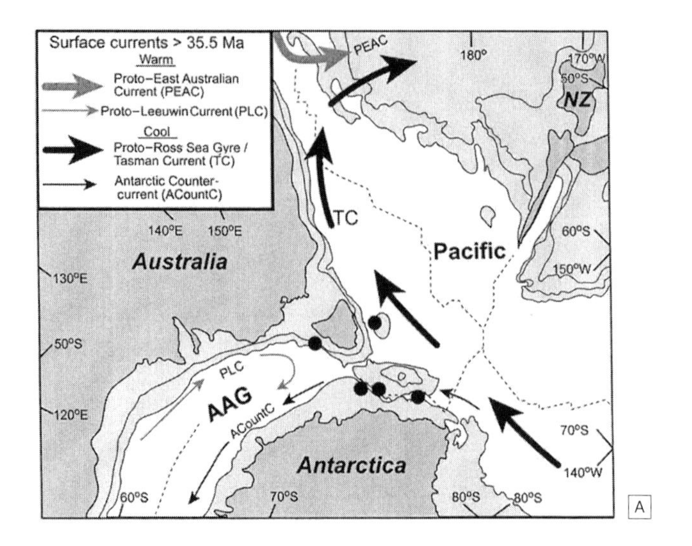

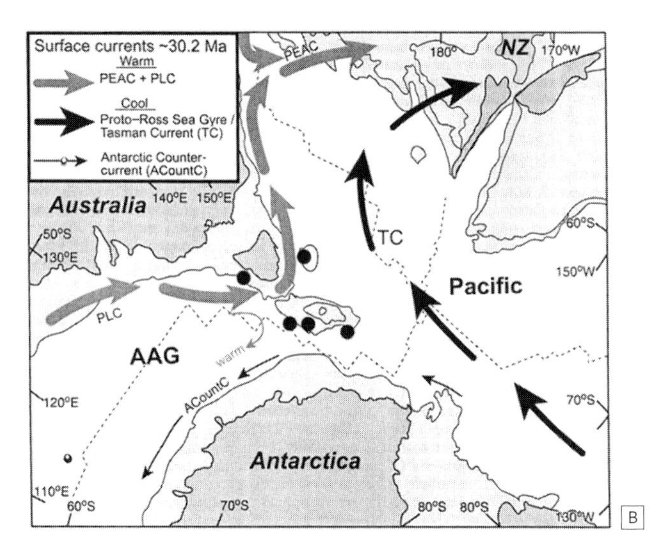

FIGURE 8.4 The most recent reconstructions of circulation around Australia and Antarctica during the Eocene-Oligocene transition: (*A*) in the late Eocene (35.5 million years ago), the circulation in the Australo-Antarctic Gulf (AAG) was blocked by Tasmania, and the large Tasman Current (TC) ran north from Antarctica and along the coast of Australia. A small Antarctic countercurrent (ACountC) was also present. (*B*) By the early Oligocene (30.2 million years ago), a series of shallower currents (Proto-Leeuwin Current [PLC] and Proto-East Australian Current [PEAC]) passed south of Tasmania and deflected the cold Tasman Current toward New Zealand. This configuration explains the earlier coring and micropaleontological data (Murphy and Kennett 1986; Exon et al. 2002) that suggested a full cold-water current passing through the gap south of Tasmania, but it is more consistent with the newer drilling data. Under this model, there was still no true ACC in the early Oligocene in this region. Solid dots indicate the drilling sites used in this study. (After Stickley et al. 2004: figs. 5 and 6)

climatic events. But recent evidence from the Tibetan Plateau (Dupont-Nivet, Hoorn, and Konert 2008; Garzione 2008) has shown that Tibet was beginning to reach high elevations in the late Eocene, about 38 million years ago. If this evidence is confirmed by future research, then high elevations in South Asia might have increased weathering and served as a reservoir for carbon dioxide.

Nevertheless, some striking events were going on in the continental interior of Eurasia during the middle Cenozoic. Earlier studies have shown that the drying up of the huge Paratethys Seaway that once covered much of western Asia had a big effect on climate because only tiny bodies of water such as the Black, Caspian, and Aral seas now remain. These drying events (for the most part dated to the Miocene) and the uplift of the Tibetan Plateau had a dramatic effect on Asian continental climate. The monsoonal climate intensified, the interior of the continent became much drier, and rates of erosion increased (Ramstein, Fluteau, and Besse 1997; An et al. 2001). Guillaume Dupont-Nivet and colleagues (2007) noticed a similar effect in the late Eocene and early Oligocene of the Tibetan Plateau. During the late Eocene, the large playa lakes dried up, and there are abundant indicators of a much cooler and drier climate during the Oligocene. These authors are not arguing that the uplift of the Tibetan Plateau was the sole cause of this aridification because, as I just mentioned, the evidence shows that most of the uplift occurred much later than the Oligocene. In fact, the change was so rapid and abrupt that it could not have been due to the relatively slow rates of mountain uplift. Instead, Dupont-Nivet and his coauthors (2007) argue that the relatively low-elevation (compared to today) Tibetan Plateau felt the global climatic effects of the Eocene–Oligocene cooling and drying, comparable to the effects that others (Retallack 1983; Zanazzi et al. 2007) have reported for the continental interior of North America. Dupont-Nivet and colleagues (2007) demonstrated that the key factor is age control. With their carefully calibrated studies using magnetostratigraphy, sedimentology, and isotopes, they were able to show that the aridification of the Tibetan Plateau was too rapid to be explained by the slow process of uplift, but had to be driven by global climatic changes. As we have seen, these processes can occur simultaneously, so deciding which one was the more important to driving climatic change depends for the most part on good dating.

Brian Jicha, Dave Scholl, and Dave Rea (2009) argued that there was a peak of explosive volcanism around the Pacific Rim during the Eocene–Oligocene transition. Such an increase in volcanism would have put aerosols into the stratosphere that could have contributed to the late Eocene cooling. Although the idea seems attractive, the actual spread of dates of eruption is very broad and spans most of the Oligocene, but the major cooling occurred at the end of the middle Eocene (37 Ma), when there was relatively little volcanism. In addition, the marine record shows abrupt cooling events and extinctions at 37 and 33 Ma, which is not consistent with the gradual increase in volcanism across the Eocene–Oligocene transition.

Robert Berner, Antonio Lasaga, and Robert Garrels (1983) argued that high rates of seafloor spreading released huge amounts of greenhouse gases (especially carbon dioxide) into the atmosphere. When the seafloor spreading slowed down, carbon dioxide production was reduced, and the gas was gradually absorbed by weathering in soils. There is pretty good evidence of very high rates of seafloor spreading in the early and middle Eocene. Plates were rearranged at 54 million years ago and again at 44 million years ago, producing at first rapid then slower spreading. However, for the past 15 million years, spreading rates have been increasing again, yet the global cooling trend has not been reversed. In addition, many scientists doubt that the amount of carbon dioxide or methane taken up by weathering in either the crustal uplift model or the seafloor-spreading model is sufficient to account for the loss of these greenhouse gases in the atmosphere.

DeConto and Pollard (2003) ran climatic simulations on a supercomputer, which takes into account all the major factors of the atmosphere, oceans, and ice caps. In their model, they allowed the carbon dioxide concentrations to decline, which then triggered small glaciers on the high plateaus of the Antarctic. After a carbon dioxide threshold was crossed, the ice sheets expanded rapidly and eventually coalesced into the continent-size East Antarctic Ice Sheet. Their model produced the known effects of the ice expansion over the Antarctic with only minimal roles for oceanic gateways. Mark Pagani and his colleagues (2005) also looked at another proxy in deep-sea cores for the partial pressure of carbon dioxide over this interval. Their data showed a stepwise drop in carbon dioxide in the middle to late Eocene, with several more drops during the Oligocene, until modern levels were reached by the latest Oligocene. If these data are correct, then they place carbon dioxide back in the driver's seat as the critical greenhouse gas and reduce the role of methane.

These studies do not address the suggestion that methane was a more important early Eocene gas than carbon dioxide (Sloan et al. 1992). However, methane has a relatively short residence time in Earth's atmosphere and eventually converts to carbon dioxide, so it was not a factor by the middle or late Eocene. Nor does this hypothesis address the studies by others (Sloan and Rea 1995; Pearson and Palmer 2000; Royer, Wing, et al. 2001; Royer 2003) showing that carbon dioxide may not have been as elevated during the early Eocene or as important to the climatic system as once thought. As Ellen Thomas (2008) argues, we need precise correlation between carbon dioxide signals and actual evidence of glaciation to establish that these factors are causally linked to the exclusion of other factors, and so far that has not been done.

Where Has All the Carbon Gone?

One of the questions not yet addressed by any model is how you remove from the atmosphere all the carbon (either as carbon dioxide or as methane) that caused the early Eocene supergreenhouse and lock it back in Earth's crust. We know that the previous transition from greenhouse to icehouse, which occurred in the Middle Carboniferous about 320 million years ago, was largely due to growth and entombment of the vegetation of enormous swamps, which locked millions of tons of coal (and its carbon) into Earth's crust. (Ironically, through our burning of this same coal, we are responsible for pushing the planet back into its greenhouse state.) But no similar reservoir can easily be identified for an Eocene carbon sink. Some limestones and coals were produced in the middle and late Eocene, but nowhere near the volume needed to account for all the carbon removed from the atmosphere. By contrast, the volume of Carboniferous coals is immense and almost certainly represents a huge transfer of carbon from the atmosphere to Earth's crust. Some scientists have suggested that the Eocene carbon might have been locked up in methane hydrates, but there is currently no way to test whether this is true. The new suggestion that the Tibetan Plateau arose in the late Eocene and increased weathering, which would have absorbed carbon dioxide (Dupont-Nivet, Hoorn, and Konert 2008; Garzione 2008), is intriguing, but needs to be supported by further evidence.

So whatever mechanism we try to invoke for the end of the Eocene greenhouse, there are problems. We know from the deep-sea cores that ocean currents changed dramatically, especially in the early Oligocene, so some scientists still believe that this factor is the critical one. What we lack with the oceanic gateways model is a mechanism to explain the evidence of cooling at 49 million years ago and through the rest of the middle and late Eocene. We have seen how some models (DeConto and Pollard 2003) argue that lowering atmospheric carbon dioxides explains more of the total picture, but they do not address the problems with carbon dioxide as the main greenhouse gas. And none of the ideas so far proposed tells us where all that carbon went when it left the atmosphere.

In short, there's much more to be learned about how the Eocene greenhouse world became our modern icehouse world, starting in the Oligocene. And that's fine. If all the answers were in hand and we knew everything, there wouldn't be any interesting questions for scientists to work on. It's great that we still have more exploring to do!

Further Reading

DeConto, R. M., and D. Pollard. 2003. Rapid Cenozoic glaciation of Antarctica induced by declining atmospheric CO_2. *Nature* 421:245–249.

Dupont-Nivet, G., C. Hoorn, and M. Konert. 2008. Tibetan uplift prior to the Eocene–Oligocene climatic transition: Evidence from pollen analysis of the Xining Basin. *Geology* 36:987–990.

Garzione, C. N. 2008. Surface uplift of Tibet and Cenozoic global cooling. *Geology* 36: 1003–1004.

Kennett, J. P. 1977. Cenozoic evolution of Antarctic glaciation, the Circum-Antarctic Ocean, and their impact on global paleoceanography. *Journal of Geophysical Research* 82: 3843–3860.

Kennett, J. P. 1981. *Marine Geology*. Englewood Cliffs, N.J.: Prentice Hall.

Kennett, J. P., R. E. Houtz, P. B. Andrews, A. R. Edwards, V. A. Gostin, M. Hahos, M. A. Hampton, D. G. Jenkins, S. V. Margolis, A. T. Ovenshine, and K. Perch-Nielsen. 1975. Cenozoic paleoceanography in the southwest Pacific Ocean: Antarctic glaciation and the development of the circum-Antarctic current. *Initial Reports of the Deep-Sea Drilling Project* 29:1155–1169.

Prothero, D. R. 2006. *After the Dinosaurs: The Age of Mammals*. Bloomington: Indiana University Press.

Thomas, E. 2008. Descent into the icehouse. *Geology* 36:191–192.

Warme, J. E., R. G. Douglas, and E. L. Winterer. 1981. The *Deep Sea Drilling Project: A Decade of Progress*. Society for Sedimentary Geology (SEPM) Special Publication no. 32. Tulsa, Okla.: SEPM.

Zachos, J. C., M. Pagani, L. C. Sloan, E. Thomas, and K. Billups. 2001. Trends, rhythms, and aberrations in global climate 65 Ma to present. *Science* 292:686–693.

The Nisqually Glacier on Mount Rainier: (*A*) in 1964; (*B*) in 1978, showing a major retreat over just 14 years. (Photographs by D. Crandell, courtesy U.S. Geological Survey)

9 | Once and Future Greenhouse?

> [Carl] Sagan called [Earth] a pale blue dot and noted that everything that has ever happened in all of human history has happened on that tiny pixel. All the triumphs and tragedies. All the wars. All the famines. It is our only home. And that is what is at stake—to have a future as a civilization. I believe this is a moral issue. It is our time to rise again to secure our future.
>
> —AL GORE, *AN INCONVENIENT TRUTH*

Vanishing Glaciers

August 2002. Taking a break from fieldwork along the northwest coast of the Olympic Peninsula of Washington, my family and I drove through Mount Rainier National Park. We stopped at all the usual tourist sights, got my son a T-shirt at the Paradise Visitor's Center, and headed out the southwest entrance back toward Tacoma. As we drove over the bridge over the Nisqually Valley, I stopped to take a picture of the Nisqually Glacier. To my shock and horror, we could no longer see the glacier from the bridge. All that remained was a river valley clogged with boulders and sand melted out of the snout of the glacier as it retreated. The water in the Nisqually River itself was almost white in color because it was clouded with the ground-up remains of rocks pulverized by the glacier, a substance known as "glacial flour."

I'd heard that the glaciers all over the world were retreating, but nothing brings that point home quite so dramatically as seeing it happen in your own lifetime. I vividly remember standing on that same bridge in June 1981. At that time, I was still a struggling graduate student finishing my dissertation at Columbia University, and in the fall of 1980 I had taken a part-time job teaching at Vassar College in Poughkeepsie, New York. They had hired me to teach sedimentary geology (not my specialty, but I had a solid undergrad background, and I had taught it as a grad student at Columbia), and my ability to teach paleontology as well was considered a bonus. (Like many paleontologists, I have had to teach whatever I can because there is little demand to teach paleontology anywhere these days.)

By 1980, Vassar was famous as one of the most respected of the "Seven Sisters" women's schools, and it had many famous alumni, from Edna St. Vincent Millay to Jackie Kennedy, Jane Fonda, and Meryl Streep. Vassar had gone co-ed in the late 1960s, but the balance was about 60 percent female and 40 percent male (like most universities and colleges today). As this one-year sabbatical-replacement job was winding down, I noticed that the school had a course in the catalog called "Cross-Country Geology Trip." None of the old faculty at Vassar had done that trip in years, but I volunteered to run it just for the experience and to earn a bit more money. Six Vassar students and I crammed into a van and drove all the way from New York to Seattle and back, seeing every geologic highlight from the Black Hills and Rockies to the Pacific Northwest and the northern Plains over seven exciting weeks. We camped all the way, and I cooked every meal, so we kept the costs down, and I saw more famous geology in those seven weeks than I've seen in any trip since then.

One of the most memorable moments of the trip was our visit to Mount Rainier. We had tried to charter a plane to fly us over Mount St. Helens to see the aerial view of the devastation, only one full year after the eruption. On our first try, the weather was too cloudy and rainy to fly, so we went to visit Rainier instead. There we climbed the paths out of the Paradise Visitor's Center and hiked around on the top of the huge Nisqually Glacier, narrowly avoiding falling into the giant crevasses that are found along the top surface. At the top of our hike, we realized that the weather had cleared, and we had an amazing view of the smoking crater of Mount St. Helens from the south flank of Mount Rainier, so we skedaddled down off the glacier, drove out the southwest entrance, and took a brief look off the Nisqually River Bridge at the impressive snout of the Nisqually Glacier before reaching the little airport and chartering a plane flight for an amazing aerial tour of the volcanic wasteland.

Thus, in only 21 years, from 1981 to 2002, the Nisqually Glacier had retreated so much that I could see the difference in just two visits to the same bridge (see the photographs that open this chapter). It turns out that this glacier's retreat is one of the best-documented examples of how all the glaciers around the world are melting due to global warming. We constantly hear about this phenomenon in the news now, but nothing brings that point home more vividly than seeing it for yourself by visiting the same glacier after only two decades and observing the incredible rate of change. In the case of Nisqually Glacier, geologists can map not only the historical evidence of its retreat, but also the numerous terminal moraines left by its retreat since the peak of the most recent glacial cooling 20,000 years ago (figure 9.1). When the bridge was built in 1914, the glacier reached all the way to the bridge abutments. It has been retreating about 100 meters (328 feet) nearly every decade in the past century, although during the 1940s it retreated almost 300 meters (984 feet). Between 1965 and 1992, it advanced and retreated

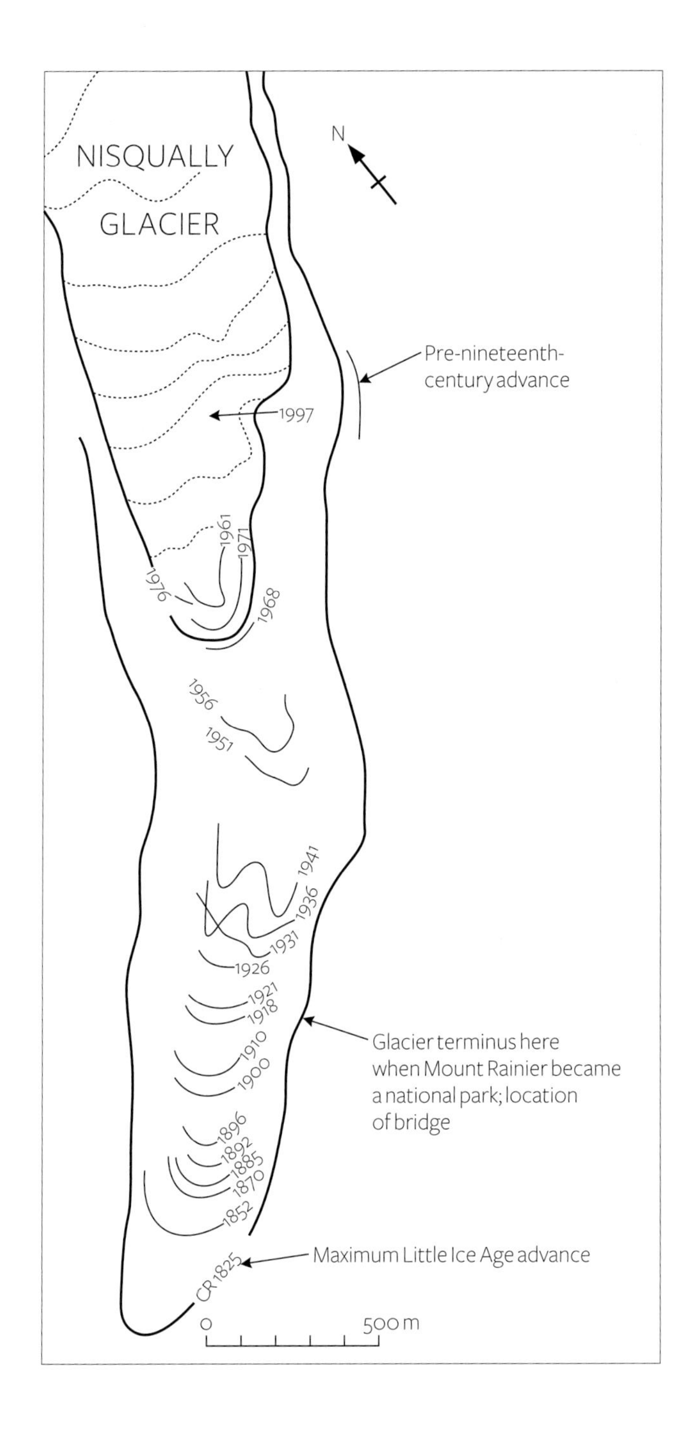

FIGURE 9.1 The retreat of the Nisqually Glacier. (Redrawn from Sigafoos and Hendricks 1961)

three times, but each time the advance was not nearly as great as the subsequent retreat, resulting in a net loss of ice. The most recent period of retreat occurred between 1985 and 1991 when the glacier thinned by 16 meters (52 feet) in the region immediately west of Glacier Vista. Today, it has retreated so far that you can no longer see the glacier from the bridge, but the snout of the glacier was clearly visible from the bridge when I first crossed it in 1981.

The same story can be said about every other glacier in the world. They all are retreating, without exception. Some are vanishing at alarming rates. Former vice president and Oscar and Nobel Prize winner Al Gore illustrates it dramatically in his documentary and book *An Inconvenient Truth* (2006). He shows example after example of "before and after" pictures of glaciers that were large less than a century ago and have now nearly vanished (figure 9.2). It is striking when mountain glaciers in the tropics (which tend to be stable in temperature during global climate change) are vanishing as well. The "snows of Kilimanjaro" given in the title of Hemingway's short story are virtually gone now, reduced to a tiny remnant block of ice that does not get large even in the wintertime. When you plot all the glacial recession rates (figure 9.3), you get a figure that shows a steady loss of ice in western North America and the Arctic, but only a slight loss of ice in Europe and the Andes (mostly because these glaciers are much smaller and do not have the mass compared to the Arctic, western North America, and central Asia).

It's far more alarming to see the vanishing of the great polar ice sheets. Every few weeks we hear another news report about the rapid shrinking of the Arctic ice cap (figure 9.4). In summer 2008, the news reported that the North Pole was ice free for the first time since the Arctic first formed about 3 million years ago (figure 9.5). Malcolm McKenna flew over the North Pole in 2000 and was one of the first humans to witness its ice breaking up. His photos were reprinted in the *New York Times*. Similarly, we see one example after another of the collapse of the Antarctic ice sheet as the floating ice on the edge of the continent breaks off rapidly. Some of the ice blocks and ice shelves that have broken up are larger than many of the states in the United States. The heart-rending images of polar bears and penguins dying out because their habitat is vanishing so rapidly are the most effective agents to remind us that the loss of the polar ice caps is not something remote that doesn't affect us, but is deadly to life at the poles.

An Inconvenient Truth and many other books and documentaries about global warming have made the public much more aware of the effects of the greenhouse gases on our planet. It is not just the obvious melting of glacial ice, but a long litany of other climatic disasters: record heat and drought in temperate and tropical parts of the world, record hurricane strengths and frequencies, extinction of some species and the spread of pest species once kept in check by cold winters, and a rise in sea level that will eventually drown most of the world's coastal cities.

FIGURE 9.2 The dramatic melting of the Muir Glacier in Glacier Bay National Park, Alaska: (*A*) photograph taken on August 13, 1941, by William O. Field; (*B*) photograph taken on August 31, 2004, by Bruce F. Molnia. ([*A*] courtesy National Snow and Ice Data Center; [*B*] courtesy U.S. Geological Survey)

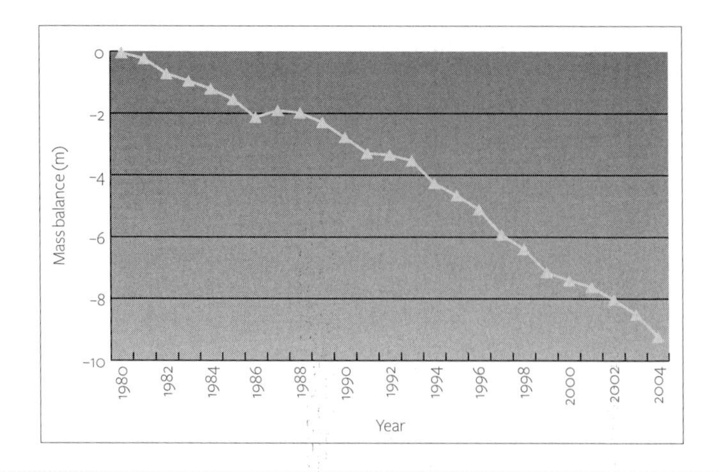

FIGURE 9.3 The mass balance loss of the world's glaciers. (Drawing modified from the World Glacier Monitoring Service)

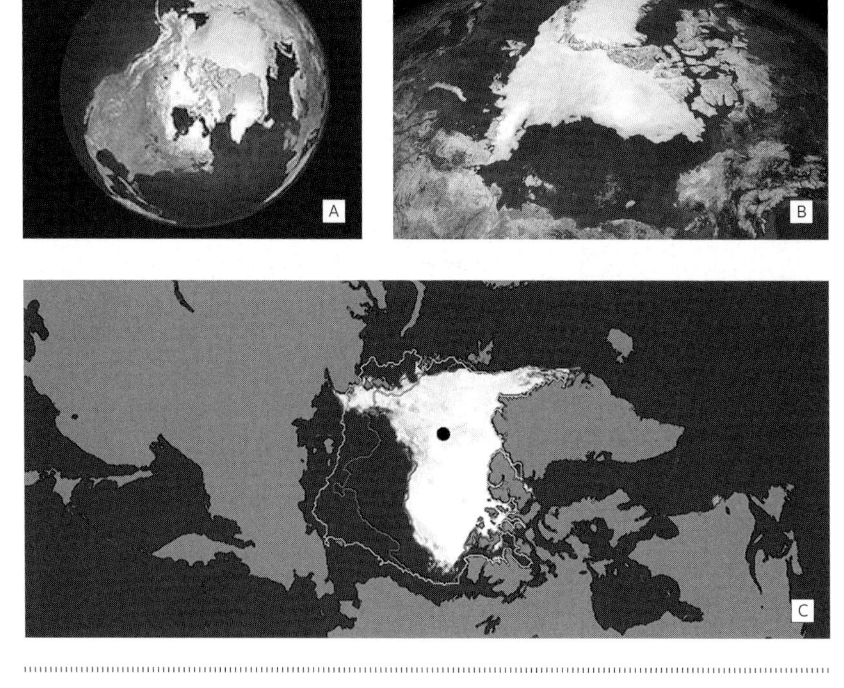

FIGURE 9.4 Images from space of the melting Arctic ice cap: (*A*) in the summer of 1979, most of the Arctic was still covered; (*B*) by the late summer of 2003, about 40 percent of the cover had vanished. (*C*) Diagram of the expected loss of the Arctic ice cap over the next decade. (Images courtesy NASA)

FIGURE 9.5 (*A*) The Arctic ice cap is so thin that the North Pole is ice free in late summer; (*B*) the North Pole as it looks before the ice melts in late summer. There's no place for Santa's castle here! ([*A*] image courtesy NOAA.; [*B*] image courtesy A. Revkin and the *New York Times*)

How Do We Know That Global Warming Is Real?

Thanks to the publicity those documentaries and books have garnered recently, all the polls now show that majorities of Americans and most western Europeans are now convinced that global warming is real, and most agree that we need to do something about it. This attitude shift is a tremendous change from just a decade ago, when the global warming naysayers (largely in the employ or influence of right-wing think tanks or the energy industries) tried their best to neutralize the virtually unanimous consensus of reputable climate scientists that global warming is real. As Gore demonstrates in the documentary *An Inconvenient Truth*, the Naomi Oreskes (2004) study showed that *all* of the 980 papers on the topic of climatic change published in the prestigious journal *Science* between 1993 and 2003 supported global warming as a reality. The "controversy" that the public once believed in was manufactured by the lobbyists and industries who had a vested interest in denying global warming. Their articles were published mostly in industry-funded journals and Web sites, not in the foremost peer-reviewed science journals—because they could not pass the muster of the finest science we are now producing.

An overwhelming scientific consensus is apparently not enough, however, for people who still think that global warming is a myth cooked up as a Communist conspiracy or who don't want to hear something that might force them to change their wasteful lifestyles. The most common counterargument one hears is that the warming is "just part of a natural cycle." This explanation sounds plausible when you see the variability of climatic curves—that is, until you look closer. The long ice cores that have come from Vostok station in Antarctica span more than 400,000 years, or four complete glacial–interglacial cycles of about 120,000 years in duration (Barnola, Bender, and Sowers 1991). The European Project for Ice Coring in Antarctica (EPICA) core from Dome C in Dronning Maud Land, Antarctica (Siegenthaler et al. 2005: Spahni et al. 2005) spans more than 650,000 years over six glacial–interglacial cycles (figure 9.6). Thus, we can see from these cores the normal range of temperature variation over many global cooling and warming cycles, and how much variability can be attributed to normal climatic change. The key point is that *at no time* in the past 650,000 years did the carbon dioxide levels exceed 300 ppm, even during the warmest interglacials. Yet our planet is already close to 400 ppm right now and is predicted to reach 600 ppm within a few decades based on the rapidly accelerating increase in carbon dioxide levels that have been measured. That increase is *not* within the range of "climatic variability"!

If starving and drowning polar bears and penguins or the evidence from long ice cores or overwhelming scientific consensus is not proof enough for you, just

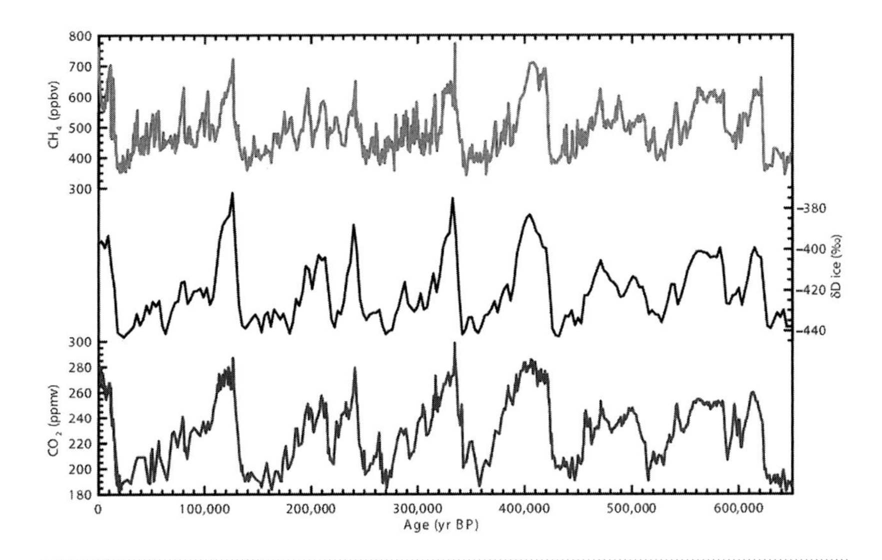

FIGURE 9.6 Six glacial–interglacial cycles and 650,000 years measured in the EPICA Antarctic ice core. ppmv = parts per million by volume; ppbv = parts per billion by volume. (Modified from Siegenthaler et al. 2005)

look at what organizations such as oil companies, the Defense Department, and insurance companies are doing now: they are responding to the situation with actions and dollars. Oil companies may still pay their lobbyists and public-relations firms to try to cloud the issue and deny the evidence of global warming, but lately they've all been spending much more time researching and investing in alternative, nonpolluting sources of energy. Rather than get caught flat-footed when cheap oil runs out in the next decade or so, they are doing the smart thing and investing in the next technology that will ensure their corporate survival. It's telling when even an oil giant such as BP, British Petroleum, now runs ads where "BP" stands for "Beyond Petroleum." The Pentagon is also taking climate change seriously and is planning accordingly.

The same story can be told about insurance companies. They have no political axe to grind with regard to the climate, and they tend to be relatively conservative about political trends and reports of future disasters. They are concerned only about properly assessing risks and figuring out the probabilities and costs associated with certain potential disasters, not about proving whether one side or the other is right for political reasons. However, they are now investing heavily in ways of insuring against disasters caused by global warming. In the alarming nature of their warnings, they even sound a bit like the climate scientists. For example, in November 2005 a Harvard–United Nations study by sponsored the global reinsurance company Swiss Re said, "Climate change will significantly affect the health of humans and ecosystems and these impacts will have economic consequences" (Epstein and Mills 2005). The report included ten case studies that outlined the current effects of climate change, from the spread of infectious diseases such as malaria and West Nile virus to extreme weather events such as heat waves and floods. The report also considered changes to forests, agriculture, marine habitat, and water. In short, the people who have to make hard business decisions are taking global warming and its consequences seriously. It seems silly when right-wing ideologues, long advocates of big business, don't take global warming as seriously as does big business itself.

Greenhouse and Icehouse in the Long Term

I have already explored how the late Mesozoic and early Cenozoic (middle Jurassic to Eocene) were a period of global greenhouse climate, with high carbon dioxide values in the atmosphere, no global ice caps, and high sea levels drowning the continents as the melted ice increased the volume of ocean water. We have seen how this greenhouse world in the late Eocene was replaced by an icehouse world by the early Oligocene, with polar ice caps growing on both Antarctica and Greenland. By 15 million years ago, the Antarctic ice cap was a permanent feature, and the complete freezing of the Arctic ice cap has apparently been around at least 3.5 million years. Thus, our "ice ages" of the past 2 million years, with their 100,000-year cycles of glaciation and interglaciation, are the normal pattern for an icehouse world. In fact, the past 10,000 years of relative warmth and stability are simply the Holocene interglacial, and if we look at any of the past few interglacials, they all ended after about 10,000 years—so we would normally be due to return to a glacial state, but that's only if the standard factors of the orbital variation control of glaciations are considered.

The most recent greenhouse–icehouse cycle, however, is just part of an even longer-term cycle of greenhouses and icehouses (figure 9.7). Prior to the Late Jurassic beginning of the "greenhouse of the dinosaurs," the planet was in an icehouse state from the Late Carboniferous (Pennsylvanian, about 300 million years ago) through the Permian and the Triassic. This icehouse may have been triggered by the growth of huge coal swamps all over the temperate latitudes, which sequestered huge amounts of atmospheric carbon dioxide in the form of coal, which was then locked up in Earth's crust. We are now returning the carbon dioxide to the atmosphere when we burn coal for fuel. Prior to the Pennsylvanian–Triassic icehouse, there was an early Paleozoic greenhouse, when nearly all the continents were flooded with huge inland seas that were filled with marine

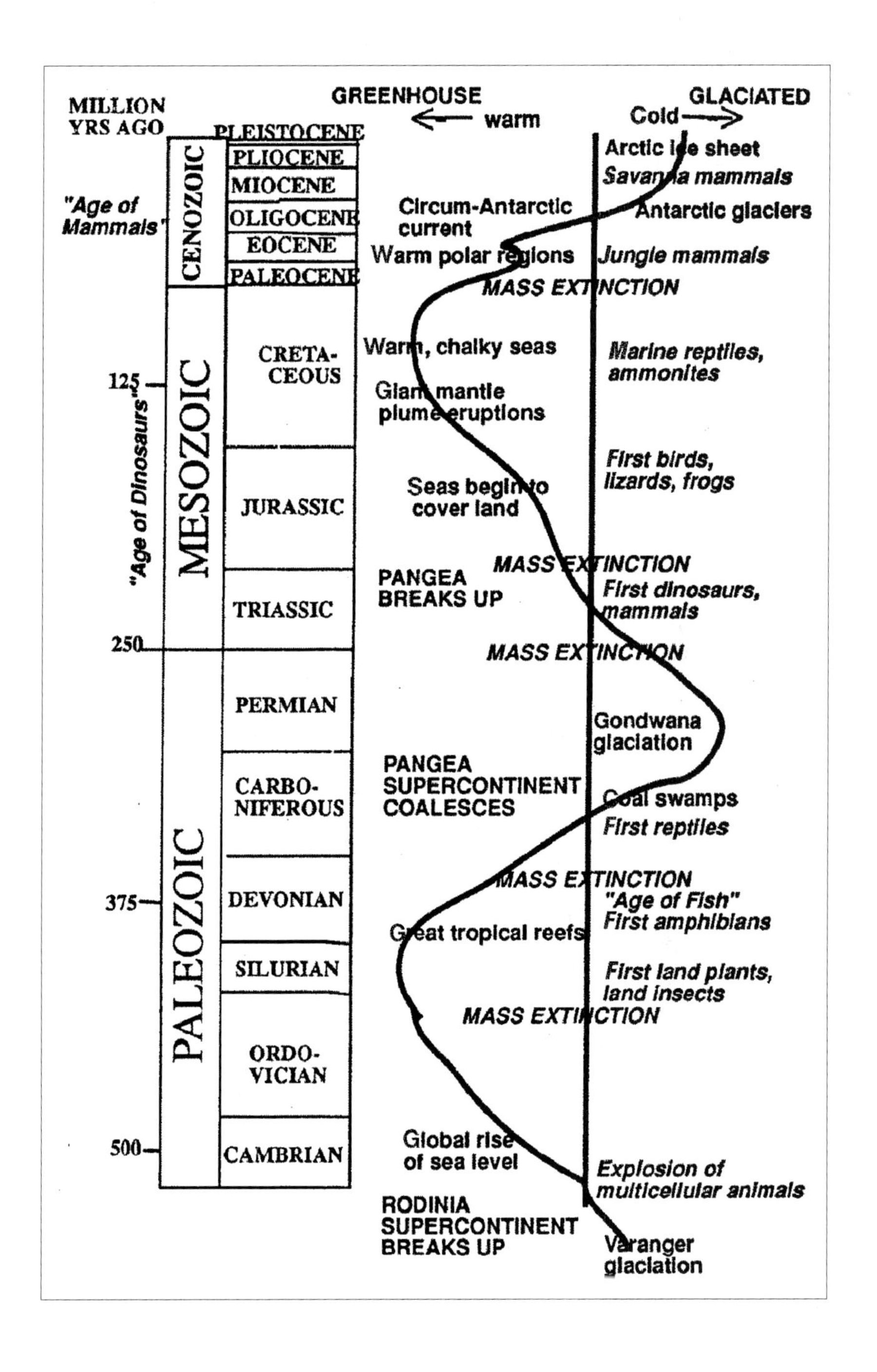

FIGURE 9.7 The past 600 million years of greenhouse–icehouse climatic cycles on Earth and the major biological and climatic events during that time.

life. You can see their fossils any time you collect in the Paleozoic beds of the U.S. Midwest or East, as I did frequently when I taught at Knox College in Illinois. Finally, the early Paleozoic greenhouse world was preceded by another event of extreme glaciation, the "snowball Earth," when the entire planet nearly froze over completely and permanently, so that there were sea-level glaciers even at the equator. This event occurred about 600 to 700 million years ago and was preceded by another warm world between then and the previous snowball Earth event around 2 billion years ago. The data on early Earth are too incomplete to tell if there was yet another greenhouse–icehouse cycle before that time, but certainly since 600 million years ago, this cycle has been one of the driving factors of global climate.

The Future?

Enough has been said and written about global warming that the possible effects are pretty familiar to most people who keep up with the news. But what the general public may not realize is two other frightening possibilities: that climate can now change much more rapidly than once thought and that global warming can actually trigger the next ice age!

The first discovery of these facts came from a series of ice cores in Greenland that had extraordinarily detailed, year-by-year records across the end of the most recent glacial and the beginning of the present interglacial (Alley 2002). Scientists found that at the end of the most recent glacial (the Younger Dryas episode, about 11,000 years ago), the region shifted from one glacial state to an interglacial state in less than a decade! This is not the overnight freeze that drove the plot of the cheesy sci-fi movie *The Day After Tomorrow*; nevertheless, a complete freeze and a mile of ice growing over the higher latitudes in a decade would be catastrophic for most of civilization. So would a rapid shift into interglacial mode in only a decade (which is not too different from what we see now with the rapid melting of ice around the world). This extremely rapid shift between one climatic mode and the other was completely unexpected among climate scientists who were used to looking at long, low-resolution ice cores and deep-sea sediment cores and thinking that climatic change usually took place over 100,000 years in many slow steps. The more detailed our records are, the less we see the pattern of gradual change that was once predicted.

That discovery leads us to our next paradox: How can global warming trigger cooling? The scientists who first pondered the rapid changes at the end of the Younger Dryas soon realized that a model Wally Broecker, Dorothy Peteet, and David Rind had postulated in 1985 might actually explain the connection. Broecker is one of the most brilliant geochemists and oceanographers in the world and the man behind many crucial ideas, such as those regarding the marine terraces that first suggested the orbital variation causes of the ice ages and the ocean's geochemistry and circulation patterns. He has received all the top awards in science. Even though he has had health problems and is approaching the age of 79, his mind is still sharp, and he has seen things clearly long before other people have. I was fortunate to take a class from him on climate changes and the ice ages when I was at Lamont-Doherty Geological Observatory of Columbia University in the late 1970s. It was an amazing experience to study under such a brilliant man who knew so much about the oceans, geochemistry, and climate change, and who was often decades ahead of the rest of the profession with his insights and somewhat unconventional ideas that turned out to be right.

Broecker first discovered the evidence (Broecker, Peteet, and Rind 1985) that the ocean currents operate like a vast "conveyer belt" (figure 9.8) that is driven by changes in temperature and salinity. The Gulf Stream, for example, is a shallow current that brings warmth and moisture from the tropics all the way to the British Isles and keeps the region relatively mild. (Remember, London and Labrador are at the same latitude, but one is *much* colder than the other!) This warm Gulf Stream eventually cools as it approaches the northern reaches of the Atlantic and sinks, whereupon the cool salty water then circulates back along the bottom of the Atlantic southward, back through the Indian and Southern ocean deep waters, and

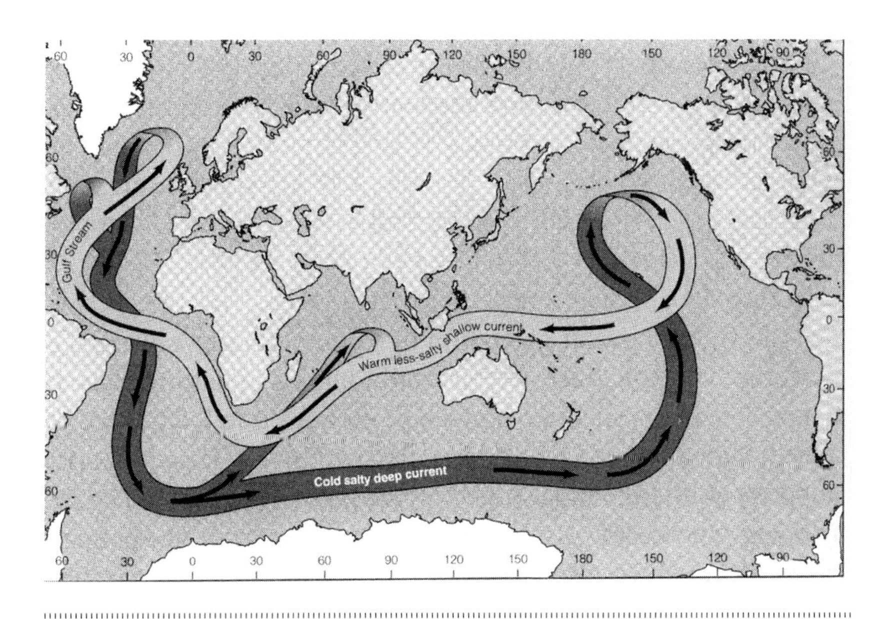

FIGURE 9.8 Wally Broecker's "conveyor belt" circulation model. (From Prothero and Dott 2003)

into the Pacific, where it finally rises again. From there, the shallow warm currents of the western Pacific circulate back through the shallow Indian Ocean, around southern Africa, and back up the surface of the Atlantic to the Gulf Stream.

As Broecker first realized, all it would take for this huge conveyer belt to be disrupted is something that changes the way the warm currents sink as they reach the northern Atlantic. During the Younger Dryas 11,000 years ago, the gradual melting as the interglacial warmed indeed triggered such a disruption. As the ice caps melted back, they released a gigantic amount of freshwater from the Gulf of St. Lawrence into the North Atlantic as various ice dams collapsed and allowed huge glacial lakes to drain catastrophically. This "lid" of freshwater shut off the thermohaline conveyer belt for more than a decade as it diverted the warm currents of the Gulf Stream much farther south and prevented their warmth and moisture from reaching from the polar regions, which had been warming and melting up to then. The Younger Dryas warming was then interrupted by a rapid increase in glaciation and then a few centuries later by an equally rapid, catastrophic warming when the conveyer belt resumed.

Are we going through something similar now? Certainly, the rapid destruction of the Arctic ice sheet is pouring huge amounts of freshwater into the North Atlantic. All it would take is a large-scale release of this water from some reservoir—say, the rapid melting of the Greenland ice cap (which is already happening as I write). Such a scenario is not implausible. Then we would have the worst of all possible scenarios—global warming to a superinterglacial greenhouse world, followed by a rapid collapse into another ice age. Either way, a stable global climate favorable to human activity, as we have been blessed with over the past few millennia, will not be in the cards.

Even faced with these unpleasant prospects of future climate, we can still have some hope. We can put our best effort into climate science so that we can refine and better predict what Earth has in store for us next. As Al Gore states in *An Inconvenient Truth*, there is a range of options we must employ to reduce our carbon footprint and slow down the trajectory to catastrophe. Eight years of George W. Bush administration obstructionism and denial of scientific reality have cost us precious time, but we still have time to take serious measures now the Obama administration has taken over. Lest we think the problem is insurmountable, we need only look at how a global effort to repair the ozone hole was successful to be reminded that the global community can make a difference when it acts together and takes the necessary steps, no matter how painful or difficult. As Al Gore puts it,

> Now it is up to use to use our democracy and our God-given ability to reason with one another about our future and make moral choices to change the policies and behaviors that would, if continued, leave a degraded, diminished and

hostile planet for our children and grandchildren—and for humankind. We must choose instead to make the 21st century a time of renewal. By seizing the opportunity that is bound up in this crisis, we can unleash the creativity, innovation, and inspiration that are just as much a part of our human birthright as our vulnerability to greed and pettiness. The choice is ours. The responsibility is ours. The future is ours. (2006:296)

Further Reading

Alley, R. 2002. *The Two-Mile Time Machine: Ice Cores, Abrupt Climate Change, and Our Future.* Princeton, N.J.: Princeton University Press.

Alley, R. 2007. Wally was right: Predictive ability of the North Atlantic "conveyer belt" hypothesis for abrupt climate change. *Annual Reviews of Earth and Planetary Science* 35:241–272.

Archer, D. 2009. *The Long Thaw: How Humans Are Changing the Next 100,000 Years of Earth's Climate.* Princeton, N.J.: Princeton University Press.

Broecker, W. S. 1997. Thermohaline circulation, the Achilles heel of our climate system: Will man-made CO_2 upset the current balance? *Science* 278:1582–1588.

Broecker, W. S. 2006. Was the Younger Dryas triggered by a flood? *Science* 312:1146–1148.

Broecker, W. S., and R. Kunzig. 2008. *Fixing Climate: What Past Climate Changes Reveal About the Current Threat—and How to Counter It.* New York: Hill and Wang.

Broecker, W. S., D. M. Peteet, and D. Rind. 1985. Does the ocean–atmosphere system have more than one stable mode of operation? *Nature* 315:21–26.

Flannery, T. 2006. *The Weather Makers: How Man Is Changing the Climate and What It Means for Life on Earth.* New York: Atlantic Monthly Press.

Gore, A. 2006. *An Inconvenient Truth: The Planetary Emergency of Global Warming and What We Can Do About It.* New York: Rodale.

Linden, E. 2006. *The Winds of Change: Climate, Weather, and the Destruction of Civilizations.* New York: Simon and Schuster.

Macdougall, J. D. 2004. *Frozen Earth: The Once and Future Story of the Ice Ages.* Berkeley: University of California Press.

Mann, M. E., and L. R. Kump. 2008. *Dire Predictions: Understanding Global Warming.* London: DK Books.

Pearce, F. 2007. *With Speed and Violence: Why Scientists Fear Tipping Points in Climate Change.* Boston: Beacon Press.

Prothero, D. R. 2006. *After the Dinosaurs: The Age of Mammals.* Bloomington: Indiana University Press.

Prothero, D. R., and R. H. Dott Jr. 2009. *Evolution of the Earth.* 8th ed. New York: McGraw-Hill.

Schneider, S. H. 1990. *Global Warming: Are We Entering the Greenhouse Century?* San Francisco: Sierra Club.

Siegenthaler, U., T. F. Stocker, E. Monnin, D. Lüthi, J. Schwander, B. Stauffer, D. Raynaud, J.-M. Barnola, H. Fischer, V. Masson-Delmotte, and J. Jouzel. 2005. Stable carbon cycle–climate relationship during the late Pleistocene. *Science* 310:1313–1317.

Paleontology may be a difficult field to get a job in, but a rainbow of rewards awaits at the end of the line—maybe even a double rainbow, like this one over Cedar Pass in Badlands National Park in 1986. (Photograph by the author)

10 | Kids, Dinosaurs, and the Future of Paleontology

> Fossil hunting is by far the most fascinating of all sports. It has some danger, enough to give it zest and probably about as much as the average modern engineered big-game hunt, and danger is wholly to the hunter. It has uncertainty and excitement and all the thrills of gambling with none of its vicious features. The hunter never knows what his bag may be, perhaps nothing, perhaps a creature never before seen by human eyes. It requires knowledge, skill, and some degree of hardihood. And its results are so much more important, more worthwhile, and more enduring than those of any other sport! The fossil hunter does not kill, he resurrects. And the result of this sport is to add to the sum of human pleasure and to the treasure of human knowledge.
>
> —GEORGE GAYLORD SIMPSON, *ATTENDING MARVELS*

As a profession, paleontology has an interesting and peculiar position in society. To the public, dinosaurs are cool, and paleontology has a positive image thanks to *Jurassic Park* and the mania for dinosaurs. Unfortunately, most of the public is so woefully ignorant of science that they confuse paleontology with archeology and think that anyone who studies ancient things must be in the same specialty. I can't count how many times I've heard the public and the media use the term *archeologist* to describe a paleontologist (and vice versa) or how many times people have asked me about ancient ruins or artifacts when I tell them I'm a paleontologist. It's the pet peeve of nearly every archeologist and paleontologist I know, and we all get annoyed at the widespread public ignorance that assumes all "ancient" things are archeology. I do try to keep up with the latest in archeology and took several classes in anthropology in college, but I don't pretend to be an expert—I'm a paleontologist and study fossil animals *other* than humans. This problem of scientific illiteracy is hard to fix, and for most adults it's too late.

To kids younger than ten, dinosaurs and paleontology are especially popular. The market for dinosaur paraphernalia is huge, and there doesn't seem to be any limit to the public fascination for dinosaurs. People have long speculated about why dinosaurs appeal to kids so much, and the simplest answer seems to be that dinosaurs are "big, fierce, and extinct." Kids love to dream and make up stories about monsters of any kind, but dinosaurs are dragons that were once real, not imaginary. In addition, kids get a sense of mastery when they learn about dinosaurs and can tell adults all about them and how to pronounce their names. Yet these scary, real monsters are no longer around, so they don't represent a real threat and don't generate nightmares the way other monsters might.

As each of my sons grows up, I find myself visiting their elementary school classrooms once a year with my box of dinosaur fossils and giving them a talk pitched to their age group. It is an amazing sight to see a whole class of kindergarteners or third graders turned on by dinosaurs and begging to ask one question after another. And I'm so glad that my sons can point to their dad with pride and brag to their classmates that "he's a paleontologist."

Whatever the reason, many kids start out as budding paleontologists and are fascinated with other aspects of science and nature as well, but something happens as they get older. By the time they reach high school, few of them are interested in science anymore, and most only take the minimum amount of science necessary to graduate. Girls in particular start out academically ahead of boys in childhood, but are sorely underrepresented in the sciences by the time they come to college. No one knows why. It may be due to cultural forces, which start to deemphasize science and learning as "uncool" when kids reach puberty. For evidence, one need look no further than cable television, which is highly niche marketed to certain age groups. The shows for preschoolers and preteens focus on learning and science, but the shows for teens are focused on pop music, fashion, and romance. If the teen channels happen to include the character of a teenager who is interested in science, he (and it's nearly always a boy, not a girl) is portrayed as an uncool, glasses-wearing nerd. From Poindexter in my childhood to Jimmy Neutron and Dexter's Laboratory today, this stereotype of science and scientists shapes our kids. Just watch a few minutes of programming on preschool channels such as Noggin or Sprout and compare them with teen channels such as Disney and MTV to see the striking differences.

Some people blame teenagers' lack of interest in science on poor science instruction in middle schools and high schools, and argue that we need to make every science class a "hands-on" interactive experiment or project rather than emphasize rote memorization. To some degree, this argument may be valid, but I've studied the books and instructional materials that my eldest son has used through middle school and high school. All in all, I've been very impressed by how well they handle subjects such as biology, physics, and chemistry. More hands-on approaches are certainly preferable, but most schools don't have the time or the resources to use them every day.

Whatever the reason, the sad result is that too many kids are turned off to science before they reach college. I teach the big introductory physical geology class (nicknamed "Rocks for Jocks" in most universities) nearly every semester, and I find that so many of my students have already decided that they have no aptitude for science and have written off science as an onerous chore. They are there just to fill their requirements. Then they take my class and are excited about geology, discovering how fun and important it is. Many people are worried for the future of the country, for we are not producing enough scientists to keep up with demand in many fields and are falling behind other Western nations in our scientific research in recent years because of this dearth.

Even more alarming, the average adult citizen has had at best only one or two indifferent science classes since elementary school, so most citizens are scientifically illiterate, as study after study shows. In science literacy, the United States typically ranks at the bottom of the list of westernized industrial countries, sitting alongside Third World countries that have a fraction of our resources and overall literacy rate. Most of the western European nations, Japan, and Canada are routinely in the top ten. As Carl Sagan put it, "We've arranged a global civilization in which the most critical elements profoundly depend on science and technology. We have also arranged things so that almost no one understands science and technology. This is a prescription for disaster" (1996:24). Or as Stephen Hawking said, "In a democracy, it is very important that the public have a basic understanding of science so that they can control the way that science and technology increasingly affect our lives" (qtd. on the Clark Foundation Web site).

Again and again, people have bemoaned our scientific illiteracy as adults and its consequences for our future. Many have wondered whether the childhood fascination for dinosaurs might be harnessed to turn out more scientists or at least more adults who are curious and literate about science. I certainly hope so, but after a few minutes of listening to teenagers talk about pop queens, rock bands, and their crushes, I'm not so optimistic. Dinosaurs remain cool for teenagers, too, but the cultural cues that hit during puberty may be even stronger.

My own story is typical in some ways and atypical in others. As I mentioned in chapter 1, I was one of those kids who got hooked on dinosaurs at age four and never really grew up. Kids today have the *Jurassic Park* and the *Land Before Time* movies, many dinosaur programs on cable television, and thousands of different dinosaur toys, games, and other paraphernalia to play with. When I grew up in the late 1950s, there was almost nothing in the way of dinosaur toys for a budding paleontologist to buy. There were just a handful of children's dinosaur books (I owned nearly all of them) and a few cheap sets of plastic dinosaur toys—that was about it. Nevertheless, I kept at my determination to become a paleontologist, while most kids lost their interest in dinosaurs (and science) and moved on to other things. When I was in the fourth grade, my teachers were so impressed with my knowledge of dinosaurs that they asked me to lecture to the sixth graders. When I reached sixth grade, my teacher, Mrs. Helene, rewarded the best boy and girl in her class with a special trip to collect fossils at Redrock Canyon in the Mojave Desert, arranged through her membership in the Natural History Museum of Los Angeles County. I was overwhelmed by the experience, although I found no fossils that day, just a lot of cool-looking concretions.

By tenth grade, I was serious about getting any experience that I could, so I volunteered for a full summer in 1970 working at the La Brea tar pits. They had just started a huge new excavation of Pit 91 (still going some 39 years later). The inexperienced volunteers like me were put to work sorting the tiny bird and rodent bones and the beetle carapaces out of the washed concentrated matrix from the tar pits. Older students, such as Mike Novacek (whom I never met that summer, but who would eventually become my friend and colleague), were allowed to work in Pit 91 and actually excavate the bones. I was only sixteen at the time and still didn't drive or own a car, so I had to manage a long bus ride from Glendale to my connecting bus on Skid Row in downtown Los Angeles, then catch the Wilshire Avenue bus all the way out to the tar pits. The ride was slow and tedious, and I actually spent more time on the bus than I did at the tar pits. Nevertheless, it was a great experience and further confirmed my career choice.

By tenth grade, I had already made my plans to get my undergraduate education at the University of California at Riverside, which had two paleontologists on staff and a strong geology and biology program, yet only 4,000 students on a campus built for 15,000. Thus, I got the advantage of a first-rate undergraduate education with small class sizes at the cheap California-resident tuition (about $200 a quarter back then). Once I got to college, I took every geology and paleontology course they offered (even the graduate-level seminars) and enough biology to finish a second major.

As I got to know more about the field of vertebrate paleontology, I was warned that it was extremely competitive, with one real job for every 10 to 50 applicants, so I was determined to work all that much harder. I took every class I could to broaden and improve my background, including chemistry, physics, philosophy, anthropology, human evolution, French and German, and enough courses in Greek and Latin that I almost finished a third major in classics. I ended up averaging six 4-unit classes every quarter (a normal load was four courses each quarter) in my four years at Riverside, which set an all-time record for total units taken in four years—a record that will never be broken because they changed the rules after I graduated.

In chapter 1, I described how I got my graduate education at Columbia University and the American Museum. Despite my first-rate education at the top program in the United States, getting a job and moving up the ranks of the profession have been a struggle. I taught at Vassar College part time in 1980 and 1981, which gave me valuable experience once I got my Ph.D. in 1982 and was ready for the job market. Despite having eight papers already published then, I received only one job offer, from Knox College in Galesburg, Illinois. In the second year I was there, my department was closed and my position eliminated. As I was getting ready to leave Knox and applying for jobs, I received one or two offers in 1985 before coming to Occidental, and I've been here ever since. But I've seen many people who are far less qualified than I am and who have published fewer papers and books than I have move up the ladder of prestigious positions in paleontology, while I've only come close in a few job searches since 1985.

One of the unfortunate things about paleontology that we don't tell the kids, but that they should know, is that the job market in paleontology is horrendous. Fewer than one in every ten students who gets a Ph.D. in the field gets a real job in paleontology. To land a decent position in this business requires tremendous dedication and perseverance, plus an attitude that there is no other choice or fallback plan: you must be determined to become a professional paleontologist or die trying. If you're not totally dedicated and willing to make whatever sacrifices you need to survive, you will not make it.

I've been to the SVP meetings every single year since 1977 (the meeting in 2009 will be my thirty-second in a row). I've also attended every single meeting of the Paleontological Society (held with the GSA meeting) since 1978, so 2009 will be my thirty-first meeting in a row for that society. As I've moved up the ranks from eager grad student to midcareer professional to one of the fifty-something established paleontologists, it has been very revealing to watch the sociological dynamics in the field. When I was a young grad student, I was eager and attended every talk, but wondered why the established scientists were a bit standoffish and hard to get to know. I soon learned that the odds of a graduate student's making it to the first job was very small (one in ten or worse), so established folks didn't want to invest too much time or energy to get to know me until I'd proven that I was going to succeed and be around for a few more years. Now I find myself one of the old guard, looking at the crowds of grad students at the SVP meeting each year and doing the same thing: letting them mingle among themselves and paying no real attention to them until they survive the brutal gauntlet of the job market.

This sad phenomenon is particularly evident in competitions designed for these students, such as the Romer Prize for the outstanding student presentation each year at the annual SVP meeting in October. When the Romer Prize was first established, it seemed to be a prestigious award, and all my cohort of graduate students worked hard to win during the early 1980s. I listen to the talk among the grad students at the SVP meeting now, and they are incredibly competitive and stressed out about the award, worrying that it will make or break their chances of becoming a professional paleontologist. But a quick glance at the list of winners (posted on the SVP Web site) and at what eventually happened to them is very surprising. Fewer than half of the winners are still in the profession, and many of the winners dropped out shortly after winning the prize. The reason is simple. Success in this profession is based on persevering, working hard, doing lots of research in grad school, and publishing it. Many people can give an impressive 15-minute talk, but it proves nothing about the skills necessary to survive and succeed. When I tried for the prize in 1981 at Ann Arbor, Michigan, my competition included such brilliant people as Dave Krause, and most of the competitors are still in the profession. The prize, though, went to a flashy, folksy talk that charmed the judges but had little scientific content, and the winner dropped out of the profession a year or so after getting her Ph.D. but not getting a job. Dave Krause and I have had the last laugh, and are proof that the award means next to nothing as a predictor of success.

Unfortunately, the crowds of students who want to get into this profession are often poorly advised, not told about the real job market or opportunities, and not given much support. (I make a point of giving a scary "reality check" talk to my prospective students early in their careers to let them know what they're up against and to make sure they aren't misled.) Many of these grad students are in tiny institutions with only one vertebrate paleontologist on staff, so they learn very little while they are there. The top institutions in vertebrate paleontology (the American Museum, Berkeley, Texas, Yale, Michigan, Florida, and just a few others) have at least two or more specialists in the subject on staff, with expertise on fish, amphibians, reptiles, birds, and mammals. These places train by far the majority of the students who are successful in finding jobs because they attract the brightest students, who are competitive and tuned in to the hottest ideas in the profession, not provincials interested only in local problems. In addition, students at the elite institutions are trained more broadly and have more specimens and research opportunities than do those in much smaller institutions. Finally, the good graduate programs encourage their students to work on multiple research projects and publish them before they defend their dissertation so that they have many different papers in print or in press before they face the brutal competition for a job.

Those students of the past few decades who have become established are now my friends because the profession is so tiny that anyone with any longevity and continuous scientific activity is soon known to the entire field. And we're talking a *small* field of active professionals—about 50 museum curators and maybe another 200 college and university professors who make up all the people whose job is research and at least part-time study of fossil vertebrates (even if they have teaching or curatorial responsibilities as well). I know nearly all of them very well because, as I mentioned, I've been to 31 SVP meetings in a row. Many are close friends, and I've collaborated with dozens of paleontologists over the years. Yet the typical SVP meeting may have 1,000 or more attendees, and the entire society has more than 2,000 members. More than half of them, however, are enthusiastic amateurs and "dino-wannabes" who find it cool to hang out with the professionals, even though most of them do not do research and do not have the graduate training. Unlike in any other professional society, amateurs make up almost 50 percent of the SVP membership and the SVP meeting attendees. There are some drawbacks to this policy, including incredibly inept, narrow, and unprofessional presentations given by some amateurs who don't realize their own incompetence and lack of qualifications. The entire profession is so tiny, however, that it could not survive on just the dues of professional vertebrate paleontologists when there are at best only 200 to 500 of them in the whole country.

If dinosaurs are so popular, why are jobs in vertebrate paleontology so scarce? Part of the problem is money. Paleontology is notoriously underfunded, and almost no paleontologist (other than superstars such as my friend, classmate, and former coauthor Paul Sereno) can claim that they get much outside grant funding. We all see the publicity and marketing surrounding dinosaurs, and think that all that money therefore must go to support paleontology. It doesn't. Paleontology is one of the least-supported areas in the entire realm of science, with only a few institutions (such as the NSF) funding its research. That's not encouraging because the odds at the NSF are so bad that less than 20 percent of the grant proposals in a given period are funded. When I was on the scientific panel to review NSF proposals one year, I was appalled by how difficult it was for any paleontologist to get grant support, even for a measly few thousand dollars. By contrast, the amount of money and the ease of funding in nearly every other area of science are much greater. Most other scientists in fields such as biology or physics or chemistry don't have to scramble hard to get funded. In my own case, all of my successful funding from the NSF and the Petroleum Research Fund of the American Chemical Society since 1985 has come to support my geophysical research in paleomagnetism; not one dollar is formally proposed or allocated for paleontological research. I do a great deal of real paleontology in the process of completing the geophysical research that I was funded to do, but I would never even consider writing a grant proposal for a purely paleontological project because it would be a waste of time.

This suggests another strategy for survival: be versatile. As Malcolm McKenna told all his students when we first arrived at the American Museum of Natural History, it's not enough to be a good vertebrate paleontologist any more and study conventional subjects. The best students also acquire an additional area of expertise or do interdisciplinary research that crosses traditional boundaries and discovers new connections between fossils and other areas of science. Malcolm himself was a pioneer in many fields, especially connecting fossil mammals to plate tectonics. Many of the students of my generation (including Bruce MacFadden, John Flynn, and me) wrote dissertations that combined paleomagnetism and paleontology. In recent years, the hottest trend has been to study the geochemisty of bones and teeth to determine ancient diets, paleophysiology, and past climatic and environmental conditions. This interdisciplinary strategy is good for several reasons. Not only does it allow you to get funded and published in ways that would never be possible if you studied only conventional vertebrate paleontology, but it also means that people can hire you as a geochemist or sedimentologist who also does vertebrate paleontology. More important, a broad cross-disciplinary approach also allows you to make breakthrough discoveries and see connections that no conventional vertebrate paleontologist could discover.

Another strike against vertebrate paleontology has to do with its position in academia. In the 1960s, when the first Baby Boom generation reached college age, the universities were expanding, and many hired lots of professors, including those in esoteric fields such as paleontology. As a late Boomer myself (born in 1954), though, I was on the bad side of this demographic shift. By the time I finished my degree in 1982, the Boomer hiring bubble had ended, and during the 1980s and 1990s most colleges and universities were cutting back. Paleontology was viewed as less essential than other fields, so when a paleontologist in a department died or retired, he was not replaced.

Paleontology also has the disadvantage of straddling the disciplines of biology and geology, so it is not firmly entrenched in either. Most paleontologists work in geology departments and teach other subjects (Earth history, sedimentary geology), including paleontology on occasion. But the grant funding and power in most geology departments are held by the geophysicists, the igneous/metamorphic petrologists, and the structural geology and tectonics specialists, most of whom seem to have a bad attitude toward paleontology. They are the ones who make the decisions about whom to hire when a paleontology slot becomes vacant. More often than not these days, they replace a true paleontologist with someone in a more trendy profession such as environmental geology, hydrogeology, or "geobiology" (which often studies only living microorganisms and never looks at fossils). As long as paleontologists can't routinely bring in six-figure research grants every year or so, they have no chance to compete and keep their positions in departments run by the almighty grant dollar. It doesn't matter that paleontology is cool and popular with the students or that "dinosaurs for jocks" courses are taught to huge numbers of students in many universities around the country. Grant dollars call the shots, and paleontologists are perpetually at a disadvantage.

Other sources of employment in vertebrate paleontology are biology and medical school anatomy programs. However, these slots are even more competitive because biology departments these days have replaced the traditional emphasis on megascopic plants and animals (organismal biology) with an emphasis on the hottest new fields with lots of grant funding, such as molecular biology, genetics, immunology, and developmental biology. I know of only a handful of my vertebrate paleontologist colleagues who hold biology department jobs, and such jobs are to be had only in the largest departments, which try to have many different biological specialties represented. Here, too, the vertebrate paleontologist is at a disadvantage because nearly every other discipline in biology has much more grant funding, and grant bucks dictate departmental research agendas.

In recent years, a number of vertebrate paleontologists have found employment in medical school anatomy departments, dental schools, and osteopathy schools. If your training and interests are much more along the lines of biology and anatomy rather than geology, this choice would make sense. Some people, however, cannot stand dissecting human cadavers year after year, so it is not for everyone. This market has dried up considerably in recent years as medical schools change and contract due to reduced demand. Twenty years ago, about half the 400 students in the frosh class at my college thought they were pre-med when they came in, but only a dozen of them actually made it to medical school. Now people are predicting a shortage of doctors in the near future as colleges see fewer pre-med students, and many medical schools are going begging for qualified applicants—all possibly due to the changes in medicine brought about by health-management organizations and the reduced financial rewards of being a doctor.

Jobs in museums are even scarcer. At best, about 50 positions for a curator of fossil vertebrates exist in the entire United States, so the competition is fierce when one or two of these jobs opens up every decade or so. And government jobs in vertebrate paleontology have just about completely vanished. The USGS let all its paleontologists go in the 1980s and 1990s, and only the Bureau of Land Management and a few other government agencies hire one or two people with paleontological expertise. Oil companies used to hire hundreds of micropaleontologists (not vertebrate paleontologists, though), but that market has almost completely vanished, despite the renewed hiring in oil companies with the rising emphasis on new exploration (Farley and Armentrout 2000).

Clearly, the handwriting is on the wall. As some researchers (Flessa and Smith 1997; Plotnick 2007a, 2007b) have showed, the number of jobs in paleontology, especially vertebrate paleontology, has been steadily declining, despite increasing numbers of students who want to get into the profession. Even more alarming is how whole institutions have simply dropped their paleo programs as someone dies or retires. The Ph.D. programs in vertebrate paleontology at a number of top-flight institutions (including Harvard, the University of Arizona, Northern Arizona University, and my alma mater, the University of California at Riverside) have vanished as their last vertebrate paleontologist retired, and they will never be revived because the job market and the lack of funding argue against retaining a paleontologist's expertise. Others, like the University of Minnesota and University of Wyoming, have replaced their traditional vertebrate paleontologist with a geochemist who also does vertebrate paleontology. As I write this, the program at Florida is threatened.

Even more serious is the loss of expertise that has come from the contraction in paleontology jobs around the country. A generation ago there were only one or two specialists for many groups of fossils. Their expertise was gathered over decades of hard work. But basic systematic paleontology is clearly unglamorous and impossible to fund, so these specialists often don't have a student who can pick up their hard-won knowledge before they die or retire. Now quite a few groups in the fossil record have no living expert to identify or study them and no one trained to take the place of the experts while they are still active.

If all this bad news does not discourage you, then you are definitely more dedicated than most students, and you may have a chance. I'm frequently asked by students all over the country—especially on the Web sites that post such advice—about how one gets into paleontology. My comments are on the Web, but here's the short version.

The first step is to recognize how difficult the journey will be and how long your odds are. You really have to ask yourself (and be really honest) whether you're willing to work that hard and sacrifice nearly everything to reach that goal. The journey starts with at least four years as a college undergrad and five to six years or more as a grad student just to earn that Ph.D. My fellow students and I used to joke that grad school is like the priesthood: the stipends are so close to starvation level and the work demands are so detrimental to your social life that you effectively take the vows of poverty and chastity. Ten years of college swallow up many prime years of your life (typically from age 18 to age 28) and a great deal of money for a profession with so little chance of success in it. By contrast, lawyers and MBAs take only 3 years to get their high-paying diplomas, and doctors and dentists a bit more. But all of these professions have substantial financial rewards at the end of the apprenticeships and a fairly strong chance that the degree holders will be employed. Paleontology guarantees neither.

If you are still determined to become a paleontologist, I would follow the advice that Reid Macdonald gave me in a wonderful three-page letter replying to a letter I wrote to him when he was the curator at the Los Angeles County Museum of Natural History. The first step is to take all the science and math classes (especially biology) that your high school offers. Work hard and get the best grades you possibly can so that you'll get into the best college your grades allow. As you look at colleges, try to focus on those that have good programs in biology or geology or both and especially those that have at least one active paleontologist on the faculty. You'll find that the list is surprisingly short because paleontologists are scarce these days, and most colleges do not have one on their faculty.

Some students prefer to go to a big university, where there is a wealth of classes in many subjects, possibly including more than one paleontology class. I personally recommend the experience of a small liberal arts college instead. At the big university, you will just be a number or a name on a roll sheet in classes of 300 to 1,000 for nearly every subject and almost never see your professor close up (let alone get to know him or her). Such universities offer few opportunities for research for undergraduates because all the research focus is on the grad students, who will actually do most of the teaching as well. Big universities are about research and grant dollars, and they don't grant tenure to professors who don't bring in the bucks and publish the papers, even if they're wonderful teachers. (I know of many specific cases of this appalling practice.)

By contrast, the small liberal arts college is teaching oriented, although many of us still do significant amounts of research as well. We don't get tenure if we aren't good teachers. In addition, the classes are tiny (nothing larger than 34 students per class in our college, and most classes have only 5 to 10 students), so you get lots of one-on-one time with the professor. Not only do the professors really get to know you, but when it comes time for recommendations, they can always write a much more detailed letter for you because they know you well. By contrast, most recommendation letters that I've read from professors at big universities are very vague and unconvincing: those professors may never have actually met the student for whom they wrote the recommendation and only know what's in their grade books.

In many small liberal arts colleges, not only do the professors do research, but they encourage their undergrads to do so as well. There are no grad students to divert their attention, so they tend to treat their junior and senior undergrads as research students instead. At my college, all of the seniors in the geology program (typically three to seven per year) are required to do a senior research project. Most start doing research with one of our five tenured faculty late in their junior year or over the following summer and come to the beginning of senior year with all their data ready to go. We then help them work through their data, formulate their hypotheses, write their abstract, and get ready to present their results at professional meetings. In the spring, they write up an extended research paper and give an oral defense.

As a consequence, most of our graduates have already had the most critical experiences needed for success in grad school (doing independent research, writing it up and presenting it, taking part in a seminar course reading primary literature), and they do very well once they get there. Grad schools love our seniors, and nearly all of them are accepted by more than one of their top choices. These same skills (gaining independent research experience, writing reports, and giving presentations) are also valuable for our students who go into environmental and groundwater geology jobs or into the oil business.

When I have lots of grant money, I spend most of it on my students. As is apparent from the field shots in this book, I take my students in the field with me to be part of my research crew, and each one of them gets a chance to publish one or more of the paleomagnetic research projects we sampled that summer—and nearly all publish their work before or soon after they graduate. In some years, I've taken several students with me on research trips to museums, so they could see all the crucial specimens needed to tackle a systematic paleontology project. The students often complete and publish those projects before they graduate.

Consequently, many of my students over the past 29 years have gone on to jobs in paleontology or are on their way: John Foster from the class of 1989, now the paleontologist at the Museum of Western Colorado; Karen Whittlesey, class of 1995, now high in the ranks of ChevronTexaco; Karina Hankins, class of 1999, now in environmental geology after getting a master's in paleontology at the University of Southern California (USC); Linda Donohoo, class of 2000, now finishing her Ph.D. at the University of New Mexico after a master's at Western Washington University; Jonathan Hoffman, class of 2003, now finishing his doctorate at the University of Wyoming after a master's at the University of Florida; Jingmai O'Connor, class of 2004, now getting her doctorate at USC; Josh Ludtke, class of 2004, now earning his doctorate at the University of Calgary after obtaining a master's at San Diego State; Matt Liter, class of 2004, and Sam McManus, class of 2005, now making big bucks in oil; and Kristina Raymond, class of 2008, just starting in the master's program at East Tennessee State University. All of these students heard my scary talk about the job market early on in their time with me, but they worked hard, published multiple projects with me, and (I'm pretty confident) have ended up or will end up employed.

So despite all the gloom and doom that I've spread through this epilogue, the bottom line is that if you really love paleontology and can't imagine spending your life doing anything else, then *go for it*. That's the way I was from the beginning, and I made it. This life has its advantages, too. The pay is not great, but you get summers and Christmas vacation off to do research if you want. With my grant funding, I'm paid to travel to some of the most beautiful places on Earth and to camp, hike, and collect fossils. I choose my own research topics and work on what I want to work on, with no boss telling me what to do—that is, as long as I produce at the end. When I'm not in class, I decide what to do with my time, and most of the week I teach only part of the day. It's definitely not a nine-to-five, five-days-a-week job. Most important of all, I get the satisfaction of making new discoveries and studying fascinating fossils and geologic problems. My research and my publications are my permanent legacy to generations of paleontologists who will come after me. Our field routinely works with scientific papers and books written, and specimens found, more than a century ago, so that's a kind of immortality that few other jobs can claim.

Yes, becoming a paleontologist is a difficult task with long odds, but if you love fossils the way I do, the rewards are worth it.

Further Reading

Farley, M. B., and J. Armentrout. 2000. Fossils in the oil patch. *Geotimes* 45:15–17.

Flessa, K. W., and D. M. Smith. 1997. Paleontology in academia: Recent trends and future opportunities. In *Paleontology in the 21st Century*. Available at http://www.nhm.ac.uk/hosted_sites/paleonet/paleo21/rr/academia.html.

Plotnick, R. E. 2007a. A somewhat fuzzy snapshot of employment in paleontology in the United States. *Palaeontologica Electronica* 11, no. 1:1–3.

Plotnick, R. E. 2007b. SWOTing at paleontology. *American Paleontologist* 15, no. 4:21–23.

Bibliography

Addicott, W. O. 1981. Brief history of Cenozoic marine biostratigraphy of the Pacific Northwest. In J. M. Armentrout, ed., *Pacific Northwest Cenozoic Biostratigraphy*, 3–15. Geological Society of America Special Paper no. 184. Boulder, Colo.: Geological Society of America.

Alley, R. 2002. *The Two-Mile Time Machine: Ice Cores, Abrupt Climate Change, and Our Future.* Princeton, N.J.: Princeton University Press.

Alley, R. 2007. Wally was right: Predictive ability of the North Atlantic "conveyer belt" hypothesis for abrupt climate change. *Annual Reviews of Earth and Planetary Science* 35:241–272.

Alvarez, L. W., W. Alvarez, F. Asaro, and H. V. Michel. 1980. Extraterrestrial cause for the Cretaceous–Tertiary extinction. *Science* 208:1095–1108.

Alvarez, W., F. Asaro, H. V. Michel, and L. W. Alvarez. 1982. Iridium anomaly approximately synchronous with terminal Eocene extinctions. *Science* 216:886–888.

An, Z., J. E. Kutzbach, W. L. Prell, and S. C. Porter. 2001. Evolution of Asian monsoons and phased uplift of the Himalaya-Tibetan plateau since late Miocene times. *Nature* 411: 62–66.

Archibald, J. D. 1996. *Dinosaur Extinctions and the End of an Era: What the Fossils Say*. New York: Columbia University Press.

Archibald, J. D., and L. J. Bryant. 1990. Differential Cretaceous–Tertiary extinctions of nonmarine vertebrates: Evidence from northeastern Montana. In V. L. Sharpton and P. W. Ward, eds., *Global Catastrophes in Earth History: The Proceedings of an Interdisciplinary Conference on Impacts, Volcanism, and Mass Mortality*, 549–562. Geological Society of America Special Paper no. 247. Boulder, Colo.: Geological Society of America.

Armentrout, J. M., ed. 1981. *Pacific Northwest Cenozoic Biostratigraphy*. Geological Society of America Special Paper no. 184. Boulder, Colo.: Geological Society of America.

Asaro, F., L. W. Alvarez, W. Alvarez, and H. V. Michel. 1982. Geochemical anomalies near the Eocene/Oligocene and Permian/Triassic boundaries. In L. T. Silver and P. H. Schultz, eds., *Geological Implications of Impacts of Large Asteroids and Comets on the Earth*, 517–528.

Geological Society of America Special Paper no. 190. Boulder, Colo.: Geological Society of America.

Aubry, M.-P., S. G. Lucas, and W. A. Berggren, eds. 1998. *Late Paleocene–Early Eocene Climatic and Biotic Events in the Marine and Terrestrial Records*. New York: Columbia University Press.

Badlands Natural History Association. 1968. *Badlands: History of Badlands National Park*. Available at: http://www.nps.gov.history/history/online_books/badl/sec1.htm.

Barker, P. F., and J. Burrell. 1977. The opening of the Drake Passage. *Marine Geology* 25:15–34.

Barker, P. F., and J. Burrell. 1982. The influence upon Southern Ocean circulation sedimentation and climate of the opening of Drake Passage. In C. Craddock, ed., *Antarctic Geoscience*, 377–385. Madison: University of Wisconsin Press.

Barnola, J. M., M. Bender, and T. Sowers. 1991. *Trends '91: A Compendium of Data on Global Change*. ORNL/CDIAC-46. Oak Ridge, Tenn.: Carbon Dioxide Information Analysis Center, Oak Ridge National Laboratory.

Becker, L., R. J. Poreda, A. R. Basu, K. O. Pope, T. M. Harrison, C. Nicholson, and R. Iasky. 2004. Bedout: A possible end-Permian impact crater offshore of northwestern Australia. *Science Express Research Article*, 10.1126/science.1093925.

Becker, L., R. J. Poreda, A. G. Hunt, T. E. Bunch, and M. Rampino. 2001. Impact event at the Permian–Triassic boundary: Evidence from extraterrestrial noble gases in fullerenes. *Science* 291:1530–1533.

Behrensmeyer, A. K., N. E. Todd, R. Potts, and G. E. McBrinn. 1997. Late Pliocene faunal turnover in the Turkana Basin, Kenya and Ethiopia. *Science* 278:1589–1594.

Berggren, W. A. 1971. Tertiary boundaries and correlations. In B. M. Funnell and W. R. Riedel, eds., *The Micropaleontology of Oceans*, 693–809. Cambridge: Cambridge University Press.

Berggren, W. A. 1972. Late Pliocene–Pleistocene glaciation. *Initial Reports of the Deep-Sea Drilling Project* 12:953–963.

Berggren, W. A. 1982. Role of ocean gateways in climatic change. In W. Berger and J. C. Crowell, eds., *Climate in Earth History*, 118–285. Washington, D.C.: National Academy of Sciences.

Berggren, W. A. 2002. Review of *Magnetic Stratigraphy of the Pacific Coast Cenozoic*, by D. R. Prothero. *Palaios* 17:527–529.

Berggren, W. A., and C. D. Hollister. 1978. Plate tectonics and paleocirculation: Commotion in the ocean. *Tectonophysics* 38:11–48.

Berggren, W. A., D. V. Kent, and J. J. Flynn. 1985. Paleogene geochronology and chronostratigraphy. *Geological Society of London Memoir* 10:141–195.

Berggren, W. A., D. V. Kent, C. C. Swisher III, and M.-P. Aubry. 1995. A revised Cenozoic geochronology and chronostratigraphy. In W. A. Berggren, D.V. Kent, C.C. Swisher III, M.-P. Aubry, and J. Hardenbol, eds., *Geochronology, Time Scales, and Global Stratigraphic Correlation*, 129–212. Society for Sedimentary Geology (SEPM) Special Publication no. 54. Tulsa, Okla.: SEPM.

Berner, R. A., D. J. Beerling, R. Dudley, J. M. Robinson, and R. A. Wildman Jr. 2003. Phanerozoic atmospheric oxygen. *Annual Reviews of Earth and Planetary Sciences* 31:105–134.

Berner, R. A., A. C. Lasaga, and R. M. Garrels. 1983. The carbonate–silicate geochemical cycle and its effect on atmospheric carbon dioxide over the past 100 million years. *American Journal of Science* 283:641–683.

Berry, W. B. N. 1999. Stratigraphic paleontology: From oil patch to academia. In E. M. Moores, D. Sloan, and D. L. Stout, eds., *Classic Cordilleran Concepts: A View from California*, 267–271. Geological Society of America Special Paper no. 338. Boulder, Colo.: Geological Society of America.

Birkelund, T., and E. Hakansson. 1982. The terminal Cretaceous in Boreal shelf seas: A multicausal event. In L. T. Silver and P. H. Schultz, eds., *Geological Implications of Impacts of Large Asteroids and Comets on the Earth*, 373–384. Geological Society of America Special Paper no. 190. Boulder, Colo.: Geological Society of America.

Birkenmajer, K. 1987. Tertiary glacial and interglacial deposits, South Shetland Islands, Antarctica: Geochronology versus biostratigraphy. *Bulletin of the Polish Academy of Science, Earth Science* 36:133–145.

Birkenmajer, K., A. Gazdzicki, K. P. Krajewski, A. Przybycin, A. Solecki, A. Tatur, and H. I. Yoon. 2005. First Cenozoic glaciers in West Antarctica. *Polish Polar Research* 26:3–12.

Boisserie, J.-R. 2005. The phylogeny and taxonomy of Hippopotamidae (Mammalia: Artiodactyla): A review based on morphology and cladistic analysis. *Zoological Journal of the Linnean Society* 143:1–26.

Boisserie, J.-R. 2007. Family Hippopotamidae. In D. R. Prothero and S. Foss, eds., *The Evolution of Artiodactyls*, 106–119. Baltimore: Johns Hopkins University Press.

Bowler, P. 1985. *Eclipse of Darwinism: Anti-Darwinian Evolutionary Theories in the Decades Around 1900*. Baltimore: Johns Hopkins University Press.

Broecker, W. S. 1987. The biggest chill. *Natural History* 96, no. 10:74–82.

Broecker, W. S. 1997. Thermohaline circulation, the Achilles heel of our climate system: Will man-made CO_2 upset the current balance? *Science* 278:1582–1588.

Broecker, W. S. 1999a. Abrupt climate change: Causal constraints provided by the paleoclimate record. *Earth Science Reviews* 51:137–154.

Broecker, W. S. 1999b. What if the conveyor were to shut down? Reflections on a possible outcome of the great global experiment. *GSA Today* 9, no. 1:1–7.

Broecker, W. S. 2006. Was the Younger Dryas triggered by a flood? *Science* 312:1146–1148.

Broecker, W. S., D. M. Peteet, and D. Rind. 1985. Does the ocean–atmosphere system have more than one stable mode of operation? *Nature* 315:21–26.

Browne, M. W. 1985. Dinosaur experts resist meteor extinction idea. *New York Times*, October 29.

Browne, M. W. 1988. The debate over dinosaur extinctions takes an unusually rancorous turn. *New York Times*, January 19.

Bryan, J. R., and D. S. Jones. 1989. Fabric of the Cretaceous–Tertiary marine macrofaunal transition at Braggs, Alabama. *Palaeogeography, Palaeoclimatology, Palaeoecology* 69:279–301.

Brysse, K. 2004. Off-limits to no one: Vertebrate paleontologists and the Cretaceous–Tertiary mass extinction. Ph.D. diss., University of Alberta.

Buick, D. P., and L. C. Ivany. 2004. 100 years in the dark: Extreme longevity of Eocene bivalves from Antarctica. *Geological Society of America Bulletin* 32:921–924.

Callahan, J. E. 1971. Velocity structure and flux of the Antarctic Circumpolar Current of South Australia. *Journal of Geophysical Research* 76:5859–5870.

Camp, C. L., ed. *James Clyman, American Frontiersman, 1792–1881*. Cleveland: Clark, 1928.

Case, J. A. 1988. Paleogene floras from Seymour Island, Antarctic Peninsula. *Geological Society of America Memoir* 169:523–530.

Chaney, R. W., and E. I. Sanborn. 1933. *The Goshen Flora of Central Oregon*. Carnegie Institute of Washington Publication no. 439. Washington, D.C.: Carnegie Institute.

Charlesworth, B., R. Lande, and M. Slatkin. 1982. A Neo-Darwinian commentary on macroevolution. *Evolution* 36:474–498.

Cheetham, A. 1986. Tempo and mode of evolution in a Neogene bryozoan: Rates of morphologic change within and across species boundaries. *Paleobiology* 12:190–202.

Cheetham, A. 1987. Tempo and mode of evolution in a Neogene bryozoan: Are trends in a single morphological character misleading? *Paleobiology* 13:286–296.

Chiappe, L. M. 1995. The first 85 million years of avian evolution. *Nature* 378:349–355.

Clemens, W. A., and L. G. Nelms. 1993. Paleoecological implications of Alaskan terrestrial vertebrate fauna in latest Cretaceous time at high paleolatitudes. *Geology* 21:503–506.

Clyman, J. 1984. *Journal of a Mountain Man*. Edited by Linda M Hasselstrom. Missoula, Mont.: Mountain Press.

CoBabe, E. 1996. Leptaucheniinae. In D. R. Prothero and R. J. Emry, eds., *The Terrestrial Eocene–Oligocene Transition in North America*, 574–580. Cambridge: Cambridge University Press.

Coccioni R., D. Basso, H. Brinkhuis, S. Galeotti, S. Gardin, S. Monechi, and S. Spezzaferri. 2000. Marine biotic signals across a late Eocene impact layer at Massignano, Italy: Evidence for long-term environmental perturbations? *Terra Nova* 12:258–263.

Courtillot, V. 1999. *Evolutionary Catastrophes: The Science of Mass Extinction*. Cambridge: Cambridge University Press.

Courtillot, V., and P. R. Renne. 2003. On the ages of flood basalt events. *Comptes Rendus Geoscience* 335:113–140.

Coxall, H. K., P. A. Wilson, H. Pälike, C. H. Lear, and J. Backman. 2005. Rapid stepwise onset of Antarctic glaciation and deeper calcite compensation depth in the Pacific Ocean. *Nature* 433:53–57.

Cuppy, W. 1941. *How to Become Extinct*. Chicago: University of Chicago Press.

Cuvier, G. 1822. *Recherches sur les ossemens fossiles, où l'on rétablit les caractères de plusieurs animaux, dont les révolutions du globe ont détruit les espèces*. Paris: Dufour and d'Ocagne.

Davies, R., J. Cartwright, J. Pike, and C. Line. 2001. Early Oligocene initiation of North Atlantic Deep Water formation. *Nature* 410:917–920.

DeConto, R. M., and D. Pollard. 2003. Rapid Cenozoic glaciation of Antarctica induced by declining atmospheric CO_2. *Nature* 421:245–249.

De Girardin, E. 1936. A trip to the Bad Lands in 1849. *South Dakota Historical Review* 1:60.

Dewar, E. 2007. The taxonomic stability of large mammals in the White River Chronofauna masked their changing dietary ecology. *Journal of Vertebrate Paleontology* 27 (suppl. to no. 3): 68A.

Dickens, G. R., M. M. Castillo, and J. C. G. Walker. 1998. A blast of gas in the latest Paleocene: Simulating first-order effects of massive dissociation of oceanic methane hydrate. *Geology* 25:258–262.

Dickens, G. R., J. R. O'Neill, D. K. Rea, and R. M. Owen. 1995. Dissociation of oceanic methane hydrate as a cause of the carbon isotope excursion at the end of the Paleocene. *Paleoceanography* 10:965–971.

Dickens, G. R., C. K. Paull, and P. Wallace. 1997. Direct measurement of in situ methane quantities in a large gas-hydrate reservoir. *Nature* 385:427–428.

Diester-Haass, L., and R. Zahn. 1996. The Eocene–Oligocene transition in the Southern Ocean: History of water masses, circulation, and biological productivity inferred from high-resolution records of stable isotopes and benthic foraminiferal abundances. *Geology* 24:16–20.

Dingus, L., and T. Rowe. 1998. *The Mistaken Extinction: Dinosaur Evolution and the Origin of Birds*. New York: Freeman.

Dobzhansky, T. 1937. *Genetics and the Origin of Species*. New York: Columbia University Press.

Dodson, P., A. K. Behrensmeyer, R. T. Bakker, and J. C. McIntosh. 1980. Taphonomy and paleoecology of the dinosaur beds of the Jurassic Morrison Formation. *Paleobiology* 6: 208–232.

Dupont-Nivet, G., C. Hoorn, and M. Konert. 2008. Tibetan uplift prior to the Eocene–Oligocene climatic transition: Evidence from pollen analysis of the Xining Basin. *Geology* 36:987–990.

Dupont-Nivet, G., W. Krijgsman, C. G. Langereis, H. A. Abels, S. Dai, and X. Fang. 2007. Tibetan Plateau aridification linked to global cooling at the Eocene–Oligocene transition. *Nature* 445:635–638.

Durden, C. J. 1966. Oligocene lake deposits in central Colorado and a new fossil insect locality. *Journal of Paleontology* 40:215–219.

Dutton, A., K. C. Lohmann, and W. J. Zinsmeister. 2002. Stable isotope and minor element proxies for Eocene climate of Seymour Island, Antarctica. *Paleoceanography* 17:106–114.

Eldredge, N. 1971. The allopatric model and phylogeny in Paleozoic invertebrates. *Evolution* 25:156–167.

Eldredge, N. 1985. *Time Frames: The Rethinking of Darwinian Evolution and the Theory of Punctuated Equilibria*. New York: Simon and Schuster.

Eldredge, N. 1995. *Reinventing Darwin: The Great Debate at the High Table of Evolutionary Theory*. New York: Wiley.

Eldredge, N., and S. J. Gould. 1972. Punctuated equilibria: An alternative to phyletic gradualism. In T. J. M. Schoft, ed., *Models in Paleobiology*, 82–115. San Francisco: Freeman Cooper.

Eldrett, J. S., I. C. Harding, P. A. Wilson, E. Butler, and A. P. Roberts. 2007. Continental ice in Greenland during the Eocene and Oligocene. *Nature* 446:176–179.

Emry, R. E. 2002. *Good Times in the Badlands*. San Jose, Calif.: Writer's Showcase.

Emry, R. J., P. R. Bjork, and L. S. Russell. 1987. The Chadronian, Orellan, and Whitneyan land mammal ages. In M. O. Woodburne, ed., *Cenozoic Mammals of North America: Geochronology and Biostratigraphy*, 118–152. Berkeley: University of California Press.

Epstein, P. R., and E. Mills, eds. 2005. *Climate Change Futures: Health, Ecological, and Economic Dimensions*. Cambridge, Mass., and Paris: Center for Health and the Global Environment of Harvard Medical School, United Nations Development Program, and Swiss Re.

Erwin, D. H. 2006. *Extinction: How Life on Earth Nearly Ended 250 Million Years Ago*. Princeton, N.J.: Princeton University Press.

Evanoff, E., D. R. Prothero, and R. H. Lander. 1992. Eocene–Oligocene climatic change in North America: The White River Formation near Douglas, east-central Wyoming. In D. R. Prothero and W. A. Berggren, eds., *Eocene–Oligocene Climatic and Biotic Evolution*, 116–130. Princeton, N.J.: Princeton University Press.

Evernden, J. F., D. E. Savage, G. H. Curtis, and G. T. James. 1964. Potassium-argon dates and the Cenozoic mammalian chronology of North America. *American Journal of Science* 262:145–198.

Exon, N., J. Kennett, M. Malone, H. Brinkhuis, G. Chaproniere, A. Ennyu, P. Fothergill, M. Fuller, M. Grauert, P. Hill, T. Janecek, C. Kelly, J. Latimer, K. McGonigal, S. Nees, U. Ninnemann, D. Neurnberg, S. Pekar, C. Pellaton, H. Pfuhl, C. Robert, U. Ruhl, S. Schellenberg, A. Shevenell, C. Stickley, N. Suzuki, Y. Touchard, W. Wei, and T. White. 2002. Drilling reveals climatic consequences of Tasmanian Gateway opening. *EOS* 83, no. 23:253–259.

Farley, K. A., and S. Mukhopadhyay. 2001. An extraterrestrial impact at the Permian–Triassic boundary? Comment. *Science* 293:2343.

Farley, M. B., and J. Armentrout. 2000. Fossils in the oil patch. *Geotimes* 45:15–17.

Fawcett, P. J., and M. B. E. Boslough. 2002. Climatic effects of an impact-induced equatorial debris ring. *Journal of Geophysical Research* 107 (D15):10129–10146.

Flannery, T. 2006. *The Weather Makers: How Man Is Changing the Climate and What It Means for Life on Earth*. New York: Atlantic Monthly Press.

Flessa, K. W., and D. M. Smith. 1997. Paleontology in academia: Recent trends and future opportunities. In *Paleontology in the 21st Century*. Available at http://www.nhm.ac.uk/hosted_sites/paleonet/paleo21/rr/academia.html.

Froehlich, D. J. 2002. Quo vadis *Eohippus*? The systematics and taxonomy of the early Eocene equids (Perissodactyla). *Zoological Journal of the Linnaean Society of London* 134: 141–256.

Galusha, T. 1975. Childs Frick and the Frick Collection of Fossil Mammals. *Curator* 18, no.1: 5–38.

Ganapathy, R. 1982. Evidence for a major meteorite impact on the Earth 34 million years ago: Implications for Eocene extinctions. *Science* 216:885–886.

Garzione, C. N. 2008. Surface uplift of Tibet and Cenozoic global cooling. *Geology* 36: 1003–1004.

Gilinsky, N. 1986. Species selection as a causal process. *Evolutionary Biology* 20:249–273.

Garzione, C. N. 2008. Surface uplift of Tibet and Cenozoic global cooling. *Geology* 36: 1003–1004.

Gilluly, J. 1977. American geology since 1910—a personal appraisal. *Annual Reviews of Earth and Planetary Sciences* 5:1–12.

Glass, B. P., D. L. DuBois, and R. Ganapathy. 1982. Relationship between an iridium anomaly and the North American micro-tektite layer in core RC9-58 from the Caribbean Sea. *Journal of Geophysical Research* 87:425–428.

Glen, W., ed. 1994. *Mass-Extinction Debates: How Science Works in a Crisis*. Stanford, Calif.: Stanford University Press.

Glikson, A. Y. 2004. Comment on "Bedout: A possible end-Permian impact crater off northwestern Australia." *Science* 306:613.

Gore, A. 1992. *Earth in the Balance: Ecology and the Human Spirit*. New York: Penguin.

Gore, A. 2006. *An Inconvenient Truth: The Planetary Emergency of Global Warming and What We Can Do About It*. New York: Rodale.

Gould, S. J. 1980a. Is a new and more general theory of evolution emerging? *Paleobiology* 6:119–130.

Gould, S. J. 1980b. The promise of paleobiology as a nomothetic evolutionary discipline. *Paleobiology* 6:96–118.

Gould, S. J. 1982a. Darwinism and the expansion of evolutionary theory. *Science* 216:380–387.

Gould, S. J. 1982b. The meaning of punctuated equilibrium and its role in validating a hierarchical approach to macroevolution. In R. Milkman, ed., *Perspectives on Evolution*, 83–104. Sunderland, Mass.: Sinauer.

Gould, S. J. 1983. Irrelevance, submission, and partnership: The changing roles of paleontology in Darwin's three centennials, and a modest proposal for macroevolution. In D. S. Bendall, ed., *Evolution from Molecules to Men*, 347–366. Cambridge: Cambridge University Press.

Gould, S. J. 1992. Punctuated equilibrium in fact and theory. In A. Somit and S. A. Peterson, eds., *The Dynamics of Evolution: The Punctuated Equilibrium Debate in the Natural and Social Sciences*, 54–84. Ithaca, N.Y.: Cornell University Press.

Gould, S. J. 2002. *The Structure of Evolutionary Theory*. Cambridge, Mass.: Harvard University Press.

Gould, S. J., and N. Eldredge. 1977. Punctuated equilibria: The tempo and mode of evolution reconsidered. *Paleobiology* 3:115–151.

Graham, A. 1999. *Late Cretaceous and Cenozoic History of North American Vegetation*. Washington, D.C.: Smithsonian Institution Press.

Gregory, K. M., and W. C. McIntosh. 1996. Paleoclimate and paleoelevation of the Oligocene Pitch-Pinnacle flora, Sawatch Range, Colorado. *Geological Society of America Bulletin* 108:545–561.

Gribben, J. 1982. *Future Weather and the Greenhouse Effect*. New York: Delacorte.

Gustavson, E. P. 1986. Preliminary biostratigraphy of the White River Group (Oligocene, Chadron, and Brule formations) in the vicinity of Chadron, Nebraska. *Transactions of the Nebraska Academy of Sciences* 14:7–19.

Hakansson, E., and E. Thomsen. 1979. Distribution and types of bryozoan communities at the boundary in Denmark. In T. Birkelund and R. G. Bromley, eds., *Cretaceous–Tertiary Boundary Events*, vol. 1, *The Maastrichtian and Danian of Denmark*, 78–91. Copenhagen: University of Copenhagen.

Hallam, A. 1990. The end-Triassic mass extinction event. In V. L. Sharpton and P. W. Ward, eds., *Global Catastrophes in Earth History: The Proceedings of an Interdisciplinary Conference on Impacts, Volcanism, and Mass Mortality*, 577–583. Geological Society of America Special Paper no. 247. Boulder, Colo.: Geological Society of America.

Hallam, A. 2004. *Catastrophes and Lesser Calamities: The Causes of Mass Extinctions*. Oxford: Oxford University Press.

Hallam, A., and P. B. Wignall. 1997. *Mass Extinctions and Their Aftermath*. Oxford: Oxford University Press.

Hansen, T. A. 1978. Larval dispersal and species longevity in Lower Tertiary gastropods. *Science* 199:885–887.

Hansen, T. A. 1980. Influence of larval dispersal and geographic distribution on species longevity in neogastropods. *Paleobiology* 6:193–207.

Hansen, T. A. 1988. Early Tertiary radiation of marine molluscs and the long-term effects of the Cretaceous–Tertiary extinction. *Paleobiology* 14:37–51.

Hansen, T. A. 1992. The patterns and causes of molluscan extinction across the Eocene/Oligocene boundary. In D. R. Prothero and W. A. Berggren, eds., *Eocene–Oligocene Climatic and Biotic Evolution*, 341–348. Princeton, N.J.: Princeton University Press.

Hansen, T. A., R. B. Farrand, H. A. Montgomery, H. G. Billman, and G. Blechschmidt. 1987. Sedimentology and extinction patterns across the Cretaceous–Tertiary boundary interval in East Texas. *Cretaceous Research* 8:229–252.

Hansen, T. A., B. R. Farrell, and B. Upshaw. 1993. The first 2 million years after the Cretaceous–Tertiary boundary in East Texas and paleoecology of the molluscan recovery. *Paleobiology* 19:251–265.

Haq, B. U. 1981. Paleogene paleoceanography: Early Cenozoic oceans revisited. *Oceanologica Acta* 4:71–82.

Harksen, J. C., and J. R. Macdonald. 1969. Guidebook to the major Cenozoic deposits of southwestern South Dakota. *South Dakota Geological Survey Guidebook* 2:1–103.

Hays, J. D., J. Imbrie, and N. J. Shackleton. 1976. Variations in the Earth's orbit—pacemaker of the ice ages. *Science* 194:1121–1132.

Heaton, T. H., and R. J. Emry.1996. Leptomerycidae. In D. R. Prothero and R. J. Emry, eds., *The Terrestrial Eocene–Oligocene Transition in North America*, 581–608. Cambridge: Cambridge University Press.

Hill, A. 1987. Causes of perceived faunal change in the later Neogene of East Africa. *Journal of Human Evolution* 16:583–596.

Hoffman, J. M., and D. R. Prothero. 2004. Revision of the late Oligocene dwarfed leptauchenine oreodont *Sespia* (Mammalia: Artiodactyla). *New Mexico Museum of Natural History and Science Bulletin* 26:153–162.

Holtz, T. 2007. Dinosaurs out of thin air: Phylogenetic perspectives on Ward's atmosphere/evolution hypothesis. *Journal of Vertebrate Paleontology* 27 (suppl. to no. 3): 91A.

Huber, M., H. Brinkhuis, C. Stickley, K. Döös, A. Sluijs, J. Warnaar, S. A. Schellenberg, and G. L. Williams. 2004. Eocene circulation of the southern ocean: Was Antarctica kept warm by subtropical waters? *Paleoceanography* 19:1–12.

Huber, M., and D. Nof. 2006. The ocean circulation in the Southern Hemisphere and its climatic impacts in the Eocene. *Paleogeography, Palaeoclimatology, Palaeoecology* 231:9–28.

Huber, M., and L. C. Sloan. 2001. Heat transport, deep waters, and thermal gradients: Coupled simulation of an Eocene greenhouse climate. *Geophysical Research Letters* 28:3481–3484.

Huber, M., L. C. Sloan, and C. Shellito. 2003. Early Paleogene oceans and climate: A fully coupled modeling approach using the NCAR CCSM. In S. L. Wing, P. D. Gingerich, B. Schmitz, and E. Thomas, eds., *Causes and Consequences of Globally Warm Climates in the Early Paleogene*, 25–48. Geological Society of America Special Paper no. 369. Boulder, Colo.: Geological Society of America.

Hutchison, J. H. 1982. Turtle, crocodilian, and champsosaur diversity changes in the Cenozoic of the north-central region of the western United States. *Palaeogeography, Palaeoclimatology, Palaeoecology* 37:149–164.

Ivany, L. C., K. C. Lohmann, F. Hasiuk, D. B. Blake, A. Glass, R. B. Aronson, and R. M. Moody. 2008. Eocene climatic record of a high southern latitude continental shelf: Seymour Island, Antarctica. *Geological Society of America Bulletin* 120:659–678.

Ivany, L. C., K. C. Lohmann, and W. P. Patterson. 2003. Paleogene temperature history of the U.S. Gulf Coastal Plain inferred from $\delta^{18}O$ of fossil otoliths. In D. R. Prothero, L. C. Ivany, and E. A. Nesbitt, eds., *From Greenhouse to Icehouse: The Marine Eocene–Oligocene Transition*, 232–251. New York: Columbia University Press.

Ivany, L. C., W. P. Patterson, and K. C. Lohmann. 2000. Cooler winters as a possible cause of mass extinctions at the Eocene/Oligocene boundary. *Nature* 407:887–890.

Ivany, L. C., B. H. Wilkinson, K. C. Lohmann, E. R. Johnson, B. J. McElroy, and G. J. Cohen. 2004. Intra-annual isotopic variation in *Venericardia* bivalves: Implications for early Eocene temperatures, seasonality, and salinity on the U.S. Gulf Coast. *Journal of Sedimentary Research* 74:7–19.

Jackson, J. B. C., and A. H. Cheetham. 1999. Tempo and mode of speciation in the sea. *Trends in Ecology and Evolution* 14, no. 2:72–77.

Jicha, B. R., D. W. Scholl, and D. K. Rea. 2009. Circum-Pacific arc flare-ups and global cooling near the Eocene–Oligocene boundary. *Geology* 37:303–306.

Kamp, P. J. J., D. B. Waghorn, and C. S. Nelson. 1990. Late Eocene–early Oligocene integrated isotope stratigraphy and biostratigraphy for paleoshelf sequences in southern Australia: Paleoceanographic implications. *Palaeogeography, Palaeoclimatology, Palaeoecology* 80:311–323.

Kauffman, E. G. 1988. The dynamics of marine stepwise mass extinctions. In M. A. Lamolda, E. G. Kauffman, and O. H. Walliser, eds., *Paleontology and Evolution: Extinction Events*, 57–71. Revista Española de Paleontologia, no. Extraordinario. Madrid: Sociedad Española de Paleontologia.

Keller, G. 2005. Impacts, volcanism, and mass extinction: Random coincidence or cause and effect? *Australian Journal of Earth Sciences* 52:725–757.

Kemp, E. M. 1975. Palynology of Leg 28 drill sites, Deep Sea Drilling Project. *Initial Reports of the Deep Sea Drilling Project* 28:599–623.

Kemp, E. M. 1978. Tertiary climatic evolution and vegetation history in the Southeast Indian Ocean region. *Palaeogeography, Palaeoclimatology, Palaeoecology* 24:169–208.

Kennett, J. P. 1977. Cenozoic evolution of Antarctic glaciation, the Circum-Antarctic Ocean, and their impact on global paleoceanography. *Journal of Geophysical Research* 82: 3843–3860.

Kennett, J. P. 1981. *Marine Geology*. Englewood Cliffs, N.J.: Prentice Hall.

Kennett, J. P., R. E. Burns, J. E. Andrews, M. Churkin Jr., T. A. Davies, P. Dumitrica, A. R. Edwards, J. S. Galehouse, G. H. Packham, and G. J. van der Lingen. 1972. Australian–Antarctic continental drift, palaeocirculation changes, and Oligocene deep-sea erosion. *Nature* 239:51–55.

Kennett, J. P., R. E. Houtz, P. B. Andrews, A. R. Edwards, V. A. Gostin, M. Hahos, M. A. Hampton, D. G. Jenkins, S. V. Margolis, A. T. Ovenshine, and K. Perch-Nielsen. 1975. Cenozoic paleoceanography in the Southwest Pacific Ocean: Antarctic glaciation and the development of the circum-Antarctic current. *Initial Reports of the Deep Sea Drilling Project* 29:1155–1169.

Kennett, J. P., and L. D. Stott. 1991. Abrupt deep-sea warming, paleoceanographic changes, and benthic extinctions at the end of the Palaeocene. *Nature* 353:225–229

Kielan-Jaworowska, Z., R. L. Cifelli, and Z. Luo. 2004. *Mammals from the Age of Dinosaurs: Origins, Evolution, and Structure*. New York: Columbia University Press.

King, P. B. 1977. *Evolution of North America*. Princeton, N. J.: Princeton University Press.

Kleinpell, R. M. 1938. *Miocene Stratigraphy of California*. Tulsa, Okla.: American Association of Petroleum Geologists.

Kleinpell, R. M. 1980. *Miocene Stratigraphy of California Revisited.* American Association of Petroleum Geologists Studies in Geology no. 11. Tulsa, Okla.: American Association of Petroleum Geologists.

Koeberl, C., K. A. Farley, B. Peucker-Ehrenbrink, and M. A. Sephton. 2004. Geochemistry of the end-Permian mass extinction event in Austria and Italy: No evidence of an extraterrestrial component. *Geology* 32:1053–1056.

Kosizek, J. 2003. New implications for the Cretaceous–Tertiary asteroid impact theory based on the persistence of extant tropical honeybees (Hymenoptera: Apidae). *Journal of Vertebrate Paleontology* 23 (suppl. to no. 3): 69A.

Kvenvolden, K. A. 1993. Gas hydrates: Geological perspective and global change. *Reviews of Geophysics* 31:173–187.

LaBandeira, C., and J. J. Sepkoski Jr. 1993. Insect diversity and the fossil record: Myth and reality. *Science* 261:310–315.

Lande, R. 1985. Expected time for random genetic drift of a population between stable phenotypic states. *Proceedings of the National Academy of Science* 82:7641–7645.

Lander, E. B. 1977. A review of the Oreodonta (Mammalia, Artiodactyla). Ph.D. diss., University of California, Berkeley.

Levinton, J. S. 1983. Stasis in progress: The empirical basis of macroevolution. *Annual Reviews of Ecology and Systematics* 14:103–137.

Lihoreau, F. 2003. Systématique et paléoécologie des Anthracotheriidae [Artiodactyla; Suiformes] du Mio–Pliocne de l'Ancien Monde: Implications paéobiogéographiques. Ph.D. diss., Université de Poitiers.

Linden, E. 2006. *The Winds of Change: Climate, Weather, and the Destruction of Civilizations.* New York: Simon and Schuster.

Livermore, R., C.-D. Hillenbrand, M. Meredith, and G. Eagles. 2007. Drake Passage and Cenozoic climate: An open and shut case? *Geochemistry Geophysics Geosystems* 8:1–12.

Lowrie, W., and W. Alvarez. 1980. One hundred years of geomagnetic polarity history. *Geology* 9:392–397.

Lyell, C. 1831–1833. *Principles of Geology*. London: John Murray.

Lyle, M., S. Gibbs, T. C. Moore, and D. K. Rea. 2007. Late Oligocene initiation of the circumpolar current: Evidence from the South Pacific. *Geology* 35:691–694.

Macdougall, J. D. 2004. *Frozen Earth: The Once and Future Story of the Ice Ages*. Berkeley: University of California Press.

MacGinitie, H. D. 1953. *Fossil Plants of the Florissant Beds, Colorado*. Carnegie Institute of Washington Publication, vol. 599. Washington, D.C.: Carnegie Institute.

MacLeod, K. G. 1994. Extinction of inoceramid bivalves in Maastrichtian strata of the Bay of Biscay region of France and Spain. *Journal of Paleontology* 68:1048–1066.

MacLeod, N., and G. Keller, eds. 1995. *Cretaceous–Tertiary Mass Extinctions: Biotic and Environmental Changes*. New York: Norton.

MacLeod, N., P. F. Rawson, P. L. Forey, F. T. Banner, M. K. Boudagher-Fadel, P. R. Bown, J. A. Burnett, P. Chambers, S. Culver, S. E. Evans, C. Jeffery, M. A. Kaminski, A. R. Lord, A. C. Milner, A. R. Milner, N. Morris, E. Owen, B. R. Rosen, A. B. Smith, P. D. Taylor, E. Urquhart, and J. R. Young. 1997. The Cretaceous–Tertiary biotic transition. *Journal of the Geological Society, London* 154:265–292.

Mader, B. J. 1989. The Brontotheriidae: A systematic revision and preliminary phylogeny of North American genera. In D. R. Prothero and R. M. Schoch, eds., *The Evolution of Perissodactyls*, 458–484. New York: Oxford University Press.

Mallory, V. S. 1959. *Lower Tertiary Stratigraphy of the California Coast Ranges*. Tulsa, Okla.: American Association of Petroleum Geologists.

Mayr, E. 1942. *Systematics and the Origin of Species*. New York: Columbia University Press.

Mayr, E. 1992. Speciational evolution or punctuated equilibria. In A. Somit and S. A. Peterson, eds., *The Dynamics of Evolution: The Punctuated Equilibrium Debate in the Natural and Social Sciences*, 21–53. Ithaca, N.Y.: Cornell University Press.

Mayr, E. 2001. *What Evolution Is*. New York: Basic Books.

McGhee, G. R., Jr. 1996. *The Late Devonian Mass Extinction*. New York: Columbia University Press.

McGhee, G. R., Jr. 2001. The "multiple impacts hypothesis" for mass extinction: A comparison of the late Devonian and late Eocene. *Palaeogeography, Palaeoclimatology, Palaeoecology* 176:47–58.

McKenna, M. C., and S. K. Bell. 1997. *Classification of Mammals Above the Species Level*. New York: Columbia University Press.

Meyer, H. W. 2003. *The Fossils of Florissant*. Washington, D.C.: Smithsonian Institution Press.

Mihlbachler, M. C. 2008. Species taxonomy, phylogeny, and biogeography of the Brontotheriidae (Mammalia: Perissodactyla). *Bulletin of the American Museum of Natural History* 311:1–475.

Miller, K. G. 1992. Middle Eocene to Oligocene stable isotopes, climate, and deep-water history: The Terminal Eocene Event? In D. R. Prothero and W. A. Berggren, eds., *Eocene–Oligocene Climatic and Biotic Evolution*, 160–177. Princeton, N.J.: Princeton University Press.

Miller, K. G., and R. G. Fairbanks. 1983. Evidence for Oligocene–middle Miocene abyssal circulation changes in the western North Atlantic. *Nature* 306:250–253.

Miller, K. G., and E. Thomas. 1985. Late Eocene to Oligocene benthic foraminiferal isotopic record, Site 574, equatorial Pacific. *Initial Reports of the Deep Sea Drilling Project* 85: 771–777.

Miller, K. G., and B. E. Tucholke. 1983. Development of Cenozoic abyssal circulation south of the Greenland-Scotland Ridge. In M. H. P. Bott, S. Saxov, M. Talwani, and J. Thiede, eds., *Structure and Development of the Greenland-Scotland Ridge*, 549–589. New York: Plenum Press.

Mohr, B. A. R. 1990. Eocene and Oligocene sporomorphs and dinoflagellate cysts from Leg 113 drill sites, Weddell Sea, Antarctica. *Proceedings of the Ocean Drilling Program* 113: 595–606.

Moore, R. C., C. G. Lalicker, and A. G. Fischer. 1953. *Invertebrate Fossils*. New York: McGraw-Hill.

Moran, K., J. Backman, H. Brinkhuis, S. C. Clemens, T. Cronin, G. R. Dickens, F. Eynaud, J. Gattacceca, M. Jakobsson, R. W. Jordan, M. Kaminski, J. King, N. Koc, A. Krylov, N. Martinez, J. Matthiessen, D. McInroy, T. C. Moore, J. Onodera, M. O'Regan, H. Pälike, B. Rea, D. Rio, T. Sakamoto, D. C. Smith, R. Stein, K. St. John, I. Suto, N. Suzuki, K. Takahashi, M. Watanabe, M. Yamamoto, J. Farrell, M. Frank, P. Kubik, W. Jokat, and Y. Kristoffersen. 2006. The Cenozoic paleoenvironment of the Arctic Ocean. *Nature* 441:601–605.

Murphy, M. G., and J. P. Kennett. 1986. Development of latitudinal thermal gradients during the Oligocene: Oxygen-isotope evidence from the Southwest Pacific. *Initial Reports of the Deep Sea Drilling Project* 90:1347–1360.

Myers, J. 2003. Terrestrial Eocene–Oligocene vegetation and climate in the Pacific Northwest. In D. R. Prothero, L. C. Ivany, and E. A. Nesbitt, eds., *From Greenhouse to Icehouse: The Marine Eocene–Oligocene Transition*, 171–188. New York: Columbia University Press.

Ness, G., S. Levi, and R. Couch. 1980. Marine magnetic anomaly timescales for the Cenozoic: A précis, critique, and synthesis. *Reviews of Geophysics and Space Physics* 18:753–770.

Officer, C. E., and C. L. Drake. 1983. The Cretaceous–Tertiary transition. *Science* 219: 1383–1390.

Officer, C., and J. Page. 1996. *The Great Dinosaur Extinction Controversy*. New York: Addison-Wesley.

O'Harra, C. [1920] 1976. The White River Badlands. *South Dakota School of Mines Bulletin* 13:1–181.

Olsen, P. E., D. V. Kent, H.-D. Sues, C. Koeberl, H. Huber, A. Montanari, E. C. Rainforth, S. J. Fowell, M. J. Szajna, and B. W. Hartline. 2002. Ascent of dinosaurs linked to an iridium anomaly at the Triassic–Jurassic boundary. *Science* 296:1305–1307.

Olsen, P. E., N. H. Shubin, and M. H. Ander. 1987. New early Jurassic tetrapod assemblages constrain Triassic–Jurassic tetrapod extinction event. *Science* 237:1025–1029.

Olson, E. C. 1960. Morphology, paleontology, and evolution. In S. Tax, ed., *Evolution After Darwin*, vol. 1, *The Evolution of Life*, 523–546. Chicago: University of Chicago Press.

Oppenheimer, J., and R. Boyle. 1990. *Dead Heat: The Race Against the Greenhouse Effect*. New York: Basic Books.

Oreskes, N. 2004. Beyond the ivory tower: The scientific consensus on climatic change. *Science* 306:1686.

Orr, E. L., and W. N. Orr. 1999. *Oregon Fossils*. Dubuque, Iowa: Kendall-Hunt.

Osborn, H. F. 1929. *The Titanotheres of Ancient Wyoming, Dakota, and Nebraska*. 2 vols. U.S. Geological Survey Monograph no. 55. Washington, D.C.: Department of the Interior.

Ostrom, J. M., and J. H. McIntosh. 2000. *Marsh's Dinosaurs: The Collections from Como Bluff*. New Haven, Conn.: Yale University Press.

Owen, D. D. 1852. *Report of a Geological Survey of Wisconsin, Iowa, and Minnesota; and Incidentally of a Portion of Nebraska Territory*. Philadelphia: Lippincott, Grambo.

Pagani , M., J. C. Zachos, K. H. Freeman, B. Tipple, and S. Bohaty. 2005. Marked decline in atmospheric carbon dioxide concentrations during the Paleogene. *Science* 309:600–603.

Palfy, J., J. K. Mortensen, and E. S. Carter. 2000. Timing the end-Triassic mass extinction: First on land, then on sea? *Geology* 51:171–172.

Parrish, J. M., J. T. Parrish, J. H. Hutchison, and R. A. Spicer. 1987. Late Cretaceous vertebrate fossils from the North Slope of Alaska and implications for dinosaur ecology. *Palaios* 2:377–389.

Pearce, F. 2007. *With Speed and Violence: Why Scientists Fear Tipping Points in Climate Change*. Boston: Beacon Press.

Pearson, P. N., I. McMillan, B. S. Wade, T. Dunkley Jones, H. K. Coxall, P. R. Bown, and C. H. Lear. 2008. Extinction and environmental change across the Eocene–Oligocene boundary in Tanzania. *Geology* 36:179–182.

Pearson, P. N., and M. R. Palmer. 1999. Middle Eocene seawater pH and atmospheric carbon dioxide concentrations. *Science* 284:1824–1826.

Pearson, P. N., and M. R. Palmer. 2000. Estimating Paleogene atmospheric pCO_2 using boron isotope analysis of foraminifera. *Geological Society of Sweden* 122:127–128.

Pearson, P. N., B. E. van Dongen, C. J. Nicholas, R. D. Pancost, S. Schouten, J. M. Singano, and B. S. Wade. 2007. Stable warm tropical climate through the Eocene epoch. *Geology* 35:211–214.

Plotnick, R. E. 2007a. A somewhat fuzzy snapshot of employment in paleontology in the United States. *Palaeontologica Electronica* 11, no. 1:1–3.

Plotnick, R. E. 2007b. SWOTing at paleontology. *American Paleontologist* 15, no. 4:21–23.

Poag, C. W. 1997. Roadblocks on the kill curve: Testing the Raup hypothesis. *Palaios* 12: 582–590.

Poag, C. W. 1999. *Chesapeake Invader*. Princeton, N.J.: Princeton University Press.

Poag, C. W., E. Mankinen, and R. D. Norris. 2003. Late Eocene impacts: Geologic record, correlation, and paleoenvironmental consequences. In D. R. Prothero, L. C. Ivany, and E. A. Nesbitt, eds., *From Greenhouse to Icehouse: The Marine Eocene–Oligocene Transition*, 495–510. New York: Columbia University Press.

Poag, C. W., D. S. Powars, L. J. Poppe, R. B. Mixon, L. E. Edwards, D. W. Folger, and S. Bruce. 1992. Deep Sea Drilling Project Site 612 bolide event: New evidence of late Eocene impacts–wave deposits and a possible impact site, U.S. East Coast. *Geology* 20:771–774.

Pomerol, C., and I. Premoli-Silva. 1986. *Terminal Eocene Events*. Amsterdam: Elsevier.

Poole, I., D. Cantrill, and D. Utescher. 2005. A multi-proxy approach to determine Antarctic terrestrial palaeoclimate during the Late Cretaceous and early Tertiary. *Palaeogeography, Palaeoclimatology, Palaeoecology* 222: 95–121

Poore, H. R., R. Samworth, N. J. White, S. M. Jones, and I. N. McCave. Neogene overflow of Northern Component Water at the Greenland-Scotland Ridge. *Geochemistry, Geophysics, Geosystems* 7, no. 6:22–33.

Popsichal, J. J. 1997. Calcareous nannoplankton mass extinction at the Cretaceous/Tertiary boundary: An update. In G. Ryder, D. Fastovsky, and S. Gartner, eds., *The Cretaceous–Tertiary Event and Other Catastrophes in Earth History*, 335–360. Geological Society of America Special Paper no. 307. Boulder, Colo.: Geological Society of America.

Prothero, D. R. 1981. New Jurassic mammals from Como Bluff, Wyoming, and the interrelations of the non-tribosphenic Theria. *Bulletin of the American Museum of Natural History* 167, no. 5:277–326.

Prothero, D. R. 1992. Punctuated equilibrium at twenty: A paleontological perspective. *Skeptic* 1, no. 3:38–47.

Prothero, D. R. 1994. *The Eocene–Oligocene Transition: Paradise Lost*. New York: Columbia University Press.

Prothero, D. R. 1995. Geochronology and magnetostratigraphy of Paleogene North American land mammal "ages": An update. In W. A. Berggren, D.V. Kent, C.C. Swisher III, M.-P. Aubry, and J. Hardenbol, eds., *Geochronology, Time Scales, and Global Stratigraphic Correlation*, 305–315. Society for Sedimentary Geology (SEPM) Special Publication no. 54. Tulsa, Okla.

Prothero, D. R. 1996a. Camelidae. In D. R. Prothero and R. J. Emry, eds., *The Terrestrial Eocene–Oligocene Transition in North America*, 591–633. Cambridge: Cambridge University Press.

Prothero, D. R. 1996b. Hyracodontidae. In D. R. Prothero and R. J. Emry, eds., *The Terrestrial Eocene–Oligocene Transition in North America*, 634–645. Cambridge: Cambridge University Press.

Prothero, D. R. 1996c. Magnetostratigraphy of the White River Group in the High Plains. In D. R. Prothero and R. J. Emry, eds., *The Terrestrial Eocene–Oligocene Transition in North America*, 247–262. Cambridge: Cambridge University Press.

Prothero, D. R. 1999. Does climatic change drive mammalian evolution? *GSA Today* 9, no. 9:1–5.

Prothero, D. R. 2001a. Chronostratigraphic calibration of the Pacific Coast Cenozoic: A summary. In D. R. Prothero, ed., *Magnetic Stratigraphy of the Pacific Coast Cenozoic*, 377–394. Society for Sedimentary Geology (SEPM), Pacific Section, Special Publication no. 91. Tulsa, Okla.: SEPM.

Prothero, D. R. 2001b. Introduction: A century of Pacific Coast Cenozoic chronostratigraphy. In D. R. Prothero, ed., *Magnetic Stratigraphy of the Pacific Coast Cenozoic*, 1–12. Society for Sedimentary Geology (SEPM), Pacific Section, Special Publication no. 91. Tulsa, Okla.: SEPM.

Prothero, D. R., ed. 2001c. *Magnetic Stratigraphy of the Pacific Coast Cenozoic*. Society for Sedimentary Geology (SEPM), Pacific Section, Special Publication no. 91. Tulsa, Okla.: SEPM.

Prothero, D. R. 2003. *Bringing Fossils to Life: An Introduction to Paleobiology*. 2d ed. Boston: McGraw-Hill.

Prothero, D. R. 2005a. Did impacts, volcanic eruptions, or climatic change affect mammalian evolution? *Palaeogeography, Palaeoclimatology, Palaeoecology* 214:283–294.

Prothero, D. R. 2005b. *The Evolution of North American Rhinoceroses*. Cambridge: Cambridge University Press.

Prothero, D. R. 2006. *After the Dinosaurs: The Age of Mammals*. Bloomington: Indiana University Press.

Prothero, D. R. 2007a. *Evolution: What the Fossils Say and Why It Matters*. New York: Columbia University Press.

Prothero, D. R. 2007b. Magnetic stratigraphy of the Eocene–Oligocene floral transition in western North America. In H. W. Meyer and D. M. Smith, eds., *Paleontology of the Upper Eocene Florissant Formation, Colorado*, 71–87. Geological Society of America Special Paper no. 435. Boulder, Colo.: Geological Society of America.

Prothero, D. R., and J. M. Armentrout. 1985. Magnetostratigraphic correlation of the Lincoln Creek Formation, Washington: Implications for the age of the Eocene–Oligocene boundary. *Geology* 13:208–211.

Prothero, D. R., and W. A. Berggren, eds. 1992. *Eocene–Oligocene Climatic and Biotic Evolution*. Princeton, N.J.: Princeton University Press.

Prothero, D. R., C. R. Denham, and H. G. Farmer. 1982. Oligocene calibration of the magnetic polarity timescale. *Geology* 10:650–653.

Prothero, D. R., C. R. Denham, and H. G. Farmer. 1983. Magnetostratigraphy of the White River Group and its implications for Oligocene geochronology. *Palaeogeography, Palaeoclimatology, Palaeoecology* 42:151–166.

Prothero, D. R., and R. H. Dott Jr. 2003. *Evolution of the Earth*. 7th ed. New York: McGraw-Hill.

Prothero, D. R., and R. J. Emry, eds. 1996. *The Terrestrial Eocene–Oligocene Transition in North America*. Cambridge: Cambridge University Press.

Prothero, D. R., and R. J. Emry. 2004. The Chadronian, Orellan, and Whitneyan North American land mammal ages. In M. O. Woodburne, ed., *Late Cretaceous and Cenozoic Mammals of North America*, 156–168. New York: Columbia University Press.

Prothero, D. R., and T. H. Heaton. 1996. Faunal stability during the early Oligocene climatic crash. *Palaeogeography, Palaeoclimatology, Palaeoecology* 127:239–256.

Prothero, D. R., L. Ivany, and E. Nesbitt. 2000. Penrose Conference report: The marine Eocene–Oligocene transition. *GSA Today* 10, no. 7:10–11.

Prothero, D. R., L. C. Ivany, and E. A. Nesbitt, eds. 2003. *From Greenhouse to Icehouse: The Marine Eocene–Oligocene Transition*. New York: Columbia University Press.

Prothero, D. R., M. Liter, L. G. Barnes, X. Wang, E. Mitchell, S. MacLeod, D. P. Whistler, R. H. Tedford, and C. E. Ray. 2008. Land mammals from the Miocene marine rocks of the Sharktooth Hill bonebed, Kern County, California. *New Mexico Museum of Natural History and Science Bulletin* 44:299–314.

Prothero, D. R., and F. Sanchez. 2008. Systematics of the leptauchenine oreodonts (Mammalia: Artiodactyla) from the Oligocene and earliest Miocene of North America. *New Mexico Museum of Natural History and Science Bulletin* 44:335–356.

Prothero, D. R., F. Sanchez, and L. L. Denke. 2008. Magnetic stratigraphy of the early to middle Miocene Olcese Sand and Round Mountain Silt, Kern County, California. *New Mexico Museum of Natural History and Science Bulletin* 44:357–364.

Prothero, D. R., and N. Shubin. 1989. The evolution of Oligocene horses. In D. R. Prothero and R. M. Schoch, eds., *The Evolution of Perissodactyls*, 142–175. New York: Oxford University Press.

Prothero, D. R., and C. C. Swisher III. 1992. Magnetostratigraphy and geochronology of the terrestrial Eocene–Oligocene transition in North America. In D. R. Prothero and W. A. Berggren, eds., *Eocene–Oligocene Climatic and Biotic Evolution*, 46–73. Princeton, N.J.: Princeton University Press.

Prothero, D. R., and M. Thompson. 2001. Magnetic stratigraphy of the type Refugian Stage (late Eocene–early Oligocene), western Santa Ynez Range, Santa Barbara County, California. In D. R. Prothero, ed., *Magnetic Stratigraphy of the Pacific Coast Cenozoic*, 119–135. Society for Sedimentary Geology (SEPM), Pacific Section, Special Publication no. 91. Tulsa, Okla.: SEPM.

Prothero, D. R., and K. E. Whittlesey. 1998. Magnetostratigraphy and biostratigraphy of the Orellan and Whitneyan land mammal "ages" in the White River Group. In D. O. Terry, H. E. LaGarry, and R. M. Hunt Jr., eds., *Depositional Environments, Lithostratigraphy, and Biostratigraphy of the White River and Arikaree Groups (Late Eocene to Early Miocene, North America)*, 39–61. Geological Society of America Special Paper no. 325. Boulder, Colo.: Geological Society of America.

Rainger, R. 2004. *An Agenda for Antiquity: Henry Fairfield Osborn and Vertebrate Paleontology at the American Museum of Natural History, 1890–1935*. Tuscaloosa: University of Alabama Press.

Rampino, M. R., and R. B. Stothers. 1988. Flood basalt volcanism during the past 250 million years. *Science* 241:663–668.

Ramstein, G., F. Fluteau, and J. Besse. 1997. Effect of orogeny, plate motion, and land–sea distribution on Eurasian climate change over the past 30 million years. *Nature* 386:788–795.

Raup, D. M. 1986. *The Nemesis Affair: A Story of the Death of the Dinosaurs and the Ways of Science*. New York: Norton.

Raup, D. M. 1991. *Extinction: Bad Genes or Bad Luck?* New York: Norton.

Raup, D. M., and J. J. Sepkoski Jr. 1984. Periodicity of extinctions in the geologic past. *Proceedings of the National Academy of Sciences* 81:801–805.

Raup, D. M., and S. Stanley. 1971. *Principles of Paleontology*. San Francisco: Freeman.

Raymo, M. E., and W. F. Ruddiman. 1992. Tectonic forcing of late Cenozoic climate. *Nature* 359:117–122.

Rea, D. K. 1992. Delivery of Himalayan sediment to the northern Indian Ocean and its relation to global climate, sea level, uplift, and seawater strontium. *Geophysical Monograph* 70:387–402.

Renne, P. R., H. J. Melosh, K. A. Farley, W. U. Reimold, C. Koeberl, M. R. Rampino, S. P. Kelly, and B. A. Ivanov. 2004. The Bedout Crater—no sign of impact. *Science* 306:610.

Retallack, G. J. 1983. *Late Eocene and Oligocene Paleosols from Badlands National Park, South Dakota*. Geological Society of America Special Paper no. 193. Boulder, Colo.: Geological Society of America.

Retallack, G. J. 2001. A 300-million-year record of atmospheric carbon dioxide from fossil plant cuticles. *Nature* 411:287–290.

Retallack, G. J., W. N. Orr, D. R. Prothero, R. A. Duncan, P. R. Kester, and C. P. Ambers. 2004. Eocene–Oligocene extinctions and paleoclimatic change near Eugene, Oregon. *Geological Society of America Bulletin* 116:817–839.

Rich, T. H., and P. V. Rich. 2000. *Dinosaurs of Darkness*. Bloomington: Indiana University Press.

Rich, T. H., P. Vickers-Rich, A. Constantine, T. Flannery, L. Kool, and N. van Klaveren. 1997. A tribosphenic mammal from the Mesozoic of Australia. *Science* 278:1438–1442.

Riggs, E. 1903. Structure and relationships of opisthocoelian dinosaurs. Part I. *Apatosaurus*. *Geological Series of the Field Columbian Museum* 82:165–196.

Robertson, D. S., M. C. McKenna, O. B. Toon, S. Hope, and J. A. Lillegraven. 2004. Survival in the first hours of the Cenozoic. *Geological Society of America Bulletin* 116:760–768.

Rosen, B. R. 2000. Algal symbiosis and the collapse and recovery of reef communities: Lazarus corals across the K-T boundary. In S. J. Culver and P. F. Rawson, eds., *Biotic Response to Global Change: The Last 145 Million Years*, 164–180. Cambridge: Cambridge University Press.

Rosen, B. R., and D. Turnsek. 1989. Extinction patterns and biogeography of scleractinian corals across the Cretaceous/Tertiary boundary. *Memoirs of the Association of Australasian Palentologists* 8:355–370.

Royer, D. L. 2003. Estimating latest Cretaceous and Tertiary atmospheric CO_2 from stomatal indices. In S. L. Wing, P. D. Gingerich, B. Schmitz, and E. Thomas, eds., *Causes and Consequences of Globally Warm Climates in the Early Paleogene*, 79–94. Geological Society of America Special Paper no. 369. Boulder, Colo.: Geological Society of America.

Royer, D. L., R. A. Berner, I. P. Montanez, N. J. Tabor, and D. J. Beerling. 2004. CO_2 as a primary driver of Phanerozoic climate change. *GSA Today* 14, no. 3:1–4.

Royer D. L., and P. Wilf. 2006. Why do toothed leaves correlate with cold climates? Gas-exchange at leaf margins provides new insights into a classic paleotemperature proxy. *International Journal of Plant Sciences* 167:11–18.

Royer, D. L., S. L. Wing, D. J. Beerling, D. W. Jolley, P. L. Koch, L. J. Hickey, and R. A. Berner. 2001. Paleobotanical evidence for near present-day levels of atmospheric CO_2 during part of the Tertiary. *Science* 292:2310–2313.

Ruddiman, W. F., and J. E. Kutzbach. 1991. Plateau uplift and climatic change. *Scientific American* 264, no. 3:66–75.

Sagan, C. 1996. *The Demon-Haunted World: Science as a Candle in the Dark.* New York: Ballantine.

Savage, D. E. 1955. *Nonmarine Lower Pliocene Sediments in California: A Geochronologic-Stratigraphic Classification.* University of California Publications in Geological Sciences no. 31. Berkeley: University of California.

Savage, D. E. 1962. Cenozoic geochronology of the fossil mammals of the Western Hemisphere. *Revista de Museo Argentina de Ciencias Naturales* 8:53–67.

Savage, D. E., and D. E. Russell. 1983. *Mammalian Paleofaunas of the World.* Reading, Mass.: Addison-Wesley.

Schankler, D. 1981. Local extinction and ecological re-entry of early Eocene mammals. *Nature* 293:135–138.

Schenck, H. G., and R. M. Kleinpell. 1936. Refugian stage of the Pacific Coast Tertiary. *American Association of Petroleum Geologists Bulletin* 20:215–255.

Schneider, S. H. 1990. *Global Warming: Are We Entering the Greenhouse Century?* San Francisco: Sierra Club.

Schopf, T. J. M., ed. 1972. *Models in Paleobiology.* San Francisco: Freeman Cooper.

Schouten, S., J. Eldrett, D. R. Greenwood, I. Harding, M. Baas, and S. Damste Jr. 2008. Onset of long-term cooling of Greenland near the Eocene–Oligocene boundary as revealed by the branched tetraether lipids. *Geology* 36:147–150.

Sclater, J. G., L. Meinke, A. Bennett, and C. Murphy. 1986. The depth of the ocean through the Neogene. *Geological Society of America Memoir* 163:1–19.

Sepkoski, J. J., Jr. 1989. Periodicity in extinction and the problem of catastrophism in the history of life. *Journal of the Geological Society of London* 146:7–19.

Sheldon, N. D., and G. J. Retallack. 2004. Regional paleoprecipitation records from the late Eocene and Oligocene of North America. *Journal of Geology* 112:487–494.

Siegenthaler, U., T. F. Stocker, E. Monnin, D. Lüthi, J. Schwander, B. Stauffer, D. Raynaud, J.-M. Barnola, H. Fischer, V. Masson-Delmonte, and J. Jouzel. 2005. Stable carbon cycle–climatic relationship during the late Pleistocene. *Science* 310:1313–1317.

Sigafoos, R. S., and E. L. Hendricks. 1961. Botanical evidence of the modern history of Nisqually Glacier, Washington. *U.S. Geological Survey Professional Paper* 387A:A1–A20.

Simpson, G. G. 1944. *Tempo and Mode of Evolution.* New York: Columbia University Press.

Simpson, G. G. [1934] 1982. *Attending Marvels: A Patagonian Journal.* Chicago: University of Chicago Press.

Skelton, P. 2003. *The Cretaceous World.* Cambridge: Cambridge University Press.

Skinner, M. F., and F. W. Johnson. 1984. Tertiary stratigraphy and the Frick Collections of fossil vertebrates from north-central Nebraska. *Bulletin of the American Museum of Natural History* 178, no. 3:215–368.

Skinner, M. F., S. M. Skinner, and R. J. Gooris. 1977. Stratigraphy and biostratigraphy of late Cenozoic deposits in central Sioux County, western Nebraska. *Bulletin of the American Museum of Natural History* 158, no. 5:263–371.

Sloan, L. C., and D. K. Rea. 1995. Atmospheric carbon dioxide and early Eocene climate: A general circulation modeling sensitivity study. *Palaeogeography, Palaeoclimatology, Palaeoecology* 119:275–292.

Sloan, L. C., J. C. G. Walker, T. C. Moore Jr., D. K. Rea, and J. C. Zachos. 1992. Possible methane-induced polar warming in the early Eocene. *Nature* 357:320–322.

Smith, A. B., and C. H. Jeffrey. 1998. Selectivity of extinction among sea urchins at the end of the Cretaceous period. *Nature* 392:69–71.

Smith, A. B., and C. H. Jeffrey. 2000. Changes in diversity, taxic composition, and life-history patterns of echinoids over the past 145 million years. In S. J. Culver and P. F. Rawson, eds., *Biotic Response to Global Change: The Last 145 Million Years*, 181–194. Cambridge: Cambridge University Press.

Somit, A., and S. A. Peterson, eds., *The Dynamics of Evolution: The Punctuated Equilibrium Debate in the Natural and Social Sciences*. Ithaca, N.Y.: Cornell University Press.

Spahni, R., J. Chappallaz, T. F. Stocker, L. Loulergue, G. Hausammann, K. Kawamura, J. Fluckiger, J. Schwander, D. Raynaud, V. Masson-Delmonte, and J. Jouzel. 2005. Atmospheric methane and nitrous oxide of the late Pleistocene from Antarctic ice cores. *Science* 310:1317–1321.

Spicer, R. A., and J. T. Parrish. 1990. Late Cretaceous–early Tertiary palaeoclimates of northern high latitudes: A quantitative view. *Journal of the Geological Society of London* 147: 329–341.

Stanley, S. M. 1975. A theory of evolution above the species level. *Proceedings of the National Academy of Sciences* 7:646–650.

Stanley, S. M. 1979. *Macroevolution: Patterns and Process*. New York: Freeman.

Stanley, S. M. 1981. *The New Evolutionary Timetable: Fossils, Genes, and the Origin of Species*. New York: Basic Books.

Stanley, S. M. 1990. Delayed recovery and the spacing of major extinctions. *Paleobiology* 16:401–414.

Stanley, S. M., and X. Yang. 1987. Approximate evolutionary stasis for bivalve morphology over millions of years: A multivariate, multilineage study. *Paleobiology* 13:113–139.

Stevens, M. S., and J. B. Stevens. 1996. Merycoidodontinae and Miniochoerinae. In D. R. Prothero and R. J. Emry, eds., *The Terrestrial Eocene–Oligocene Transition in North America*, 498–574. Cambridge: Cambridge University Press.

Stickley, C. E., H. Brinkhuis, S. A. Schellenberg, A. Sluijs, U. Röhl, M. Fuller, M. Grauert, M. Huber, J. Warnaar, and G. L. Williams. 2004. Timing and nature of the deepening of the Tasmanian Gateway. *Paleoceanography* 19:1–15.

Surlyk, F., and M. B. Johansen. 1984. End-Cretaceous brachiopod extinctions in the Chalk of Denmark. *Science* 223:1174–1177.

Tanner, L. H., S. G. Lucas, and M. G. Chapman. 2003. Assessing the record and causes of late Triassic extinctions. *Earth Science Reviews* 65:103–139.

Tedford, R. H., J. Swinehart, D. R. Prothero, C. C. Swisher III, S. A. King, and T. E. Tierney. 1996. The Whitneyan–Arikareean transition in the High Plains. In D. R. Prothero and R. J. Emry, eds., *The Terrestrial Eocene–Oligocene Transition in North America*, 295–317. Cambridge: Cambridge University Press.

Terry, D. O., H. E. LaGarry, and R. M. Hunt Jr., eds. 1998. *Depositional Environments, Lithostratigraphy, and Biostratigraphy of the White River and Arikaree Groups* (*Late Eocene to Early Miocene, North America*). Geological Society of America Special Paper no. 325. Boulder, Colo.: Geological Society of America.

Thomas, D. J., J. C. Zachos, T. J. Bralower, E. Thomas, and S. Bohaty. 2002. Warming the fuel for the fire: Evidence for thermal dissociation of methane hydrate during the Paleocene–Eocene Thermal Maximum. *Geology* 30:1067–1070.

Thomas, E. 2008. Descent into the icehouse. *Geology* 36:191–192.

Thomas, E., and N. J. Shackleton. 1996. The Paleocene–Eocene boundary foraminiferal extinction and stable isotope anomalies. *Geological Society of London Special Publication* 101:401–411.

Thomson, K. S. 1988. Anatomy of the extinction debate. *American Scientist* 76:59–61.

Tripati, A., J. Backman, H. Elderfield, and P. Ferretti. 2005. Eocene bipolar glaciation associated with global carbon cycle changes. *Nature* 436:341–346.

Tripati, A., R. A. Eagle, C. F. Dawber, A. Morton, J. A. Dowdesdwell, K. Atkinson, Y. Bahe, E. Khadun, R. M. H. Shaw, O. Shorttle, and L. Thanabalasundraran. 2008. Evidence for Northern Hemisphere glaciation back to 44 Ma form ice-rafted debris in the Greenland Sea. *Earth and Planetary Sciences Letters* 265:112–122.

Van Valkenburgh, B. 1985. Trophic diversity in the past and present guilds of large predatory mammals. *Paleobiology* 14:155–173.

Vonhof, H. B., J. Smit, H. Brinkhuis, A. Montanari, and A. J. Nederbracht. 2000. Global cooling accelerated by early–late Eocene impacts? *Geology* 28:687–690.

Vrba, E. S. 1985. Environment and evolution: Alternative causes of temporal distribution of evolutionary events. *South African Journal of Science* 81:229–236.

Vrba, E. S. 1993. Turnover-pulses, the Red Queen, and related topics. *American Journal of Science* 293A:418–452.

Vrba, E. S., and S. J. Gould. 1986. The hierarchical expansion of sorting and selection: Sorting and selection cannot be equaled. *Paleobiology* 12:217–228.

Ward, P. D. 2006. *Out of Thin Air: Dinosaurs, Birds, and Earth's Ancient Atmosphere*. Washington, D.C.: Joseph Henry Press.

Ward, P. D. 2007. *Under a Green Sky: Global Warming, the Mass Extinctions of the Past, and What They Can Tell Us About Our Future*. Washington, D.C.: Smithsonian Books.

Ward, P. D., W. J. Kennedy, and K. G. MacLeod. 1991. Ammonite and inoceramid bivalve extinction patterns in Cretaceous/Tertiary boundary sections of the Biscay region (southwestern France, northern Spain). *Geology* 19:1181–1184.

Warme, J. E., R. G. Douglas, and E. L. Winterer. 1981. The *Deep Sea Drilling Project: A Decade of Progress*. Society for Sedimentary Geology (SEPM) Special Publication no. 32. Tulsa, Okla.: SEPM.

Warren, B. A. 1971. Antarctic deep-water circulation contribution to the world ocean. *Research in the Antarctic* 93:640–643.

Weaver, C. E. 1916. *Tertiary Faunal Horizons of Western Washington*. University of Washington Publications in Geology no. 1. Seattle: University of Washington.

Weaver, C. E. 1942. *Paleontology of the Marine Tertiary Formations of Oregon and Washington*. University of Washington Publications in Geology no. 5. Seattle: University of Washington.

Weaver, C. E., R. S. Beck, M. N. Bramlette, S. A. Carlson, L. C. Forrest, F. R. Kelley, R. M. Kleinpell, W. C. Putnam, N. L. Taliaferro, R. R. Thorup, W. A. Ver Wiebe, and E. A. Watson. 1944. Correlation of the marine Cenozoic formations of western North America. *Geological Society of America Bulletin* 55:569–598.

Wei, W. 1989. Reevaluation of the Eocene ice-rafting record from subantarctic cores. *Antarctic Journal of the United States* 1989:108–109.

Weil, A. 1984. Acid rain as an agent of extinction at the K/T boundary—NOT! *Journal of Vertebrate Paleontology* 14, no. 3:51A.

Weiner, J. 1990. *The Next One Hundred Years: Shaping the Future of Our Living Earth.* New York: Bantam Books.

What will happen to geology? 1969. *Nature* 221:903.

Wignall, P. B., B. Thomas, R. Willink, and J. Watling. 2004. The Bedout crater—no sign of impact. *Science* 306:609.

Wing, S. L., P. D. Gingerich, B. Schmitz, and E. Thomas, eds. 2003. *Causes and Consequences of Globally Warm Climates in the Early Paleogene.* Geological Society of America Special Paper no. 369. Boulder, Colo.: Geological Society of America.

Wolfe, J. A. 1971. Tertiary climatic fluctuations and methods of analysis of Tertiary floras. *Palaeogeography, Palaeoclimatology, Palaeoecology* 9:27–57.

Wolfe, J. A. 1978. A paleobotanical interpretation of Tertiary climates in the Northern Hemisphere. *American Scientist* 66:694–703.

Wolfe, J. A. 1980. Tertiary climates and floristic relationships at high latitudes in the Northern Hemisphere. *Palaeogeography, Palaeoclimatology, Palaeoecology* 30:313–323.

Wolfe, J. A. 1992. Climatic, floristic, and vegetational changes near the Eocene/Oligocene boundary in North America. In D. R. Prothero and W. A. Berggren, eds., *Eocene–Oligocene Climatic and Biotic Evolution*, 421–436. Princeton, N.J.: Princeton University Press.

Wolfe, J. A. 1994. Tertiary climatic changes at middle latitudes of western North America. *Palaeogeography, Palaeoclimatology, Palaeoecology* 108:195–205.

Wolfe, J. A., and D. M. Hopkins. 1967. Climatic changes recorded by Tertiary land floras in northwestern North America. In K. Hatai, ed., *Tertiary Correlation and Climatic Changes in the Pacific*, 7–76. Tokyo: Sasaki.

Wolfe, J. A., and H. E. Schorn. 1989. Paleoecologic, paleoclimatic, and evolutionary significance of the Oligocene Creede flora, Colorado. *Paleobiology* 15:180–198.

Wood, H. E., II, R. W. Chaney, J. Clark, E. H. Colbert, G. L. Jepsen, J. B. Reeside, and C. Stock. 1941. Nomenclature and correlation of the North American continental Tertiary. *Geological Society of America Bulletin* 51:1–48.

Zachos, J. C., M. Pagani, L. C. Sloan, E. Thomas, and K. Billups. 2001. Trends, rhythms, and aberrations in global climate 65 Ma to present. *Science* 292:686–693.

Zanazzi, A., M. J. Kohn, B. J. MacFadden, and D. O. Terry Jr. 2007. Large temperature drop across the Eocene–Oligocene transition in central North America. *Nature* 445:639–642.

Zinsmeister, W. J., and R. M. Feldmann. 1993. Late Cretaceous faunal changes in the high southern latitudes: A harbinger of impending global biotic catastrophe? *Geological Society of America Abstracts with Programs* 25, no. 6:295.

Zinsmeister, W. J., R. M. Feldmann, M. O. Woodburne, and D. H. Elliott. 1989. Latest Cretaceous/earliest Tertiary transition on Seymour Island. *Journal of Paleontology* 63:731–738.

Index